T0224748

Technologien für die intelligente Automation

Technologies for Intelligent Automation

Band 13

Reihe herausgegeben von

inIT – Institut für industrielle Informationstechnik
Lemgo, Deutschland

Ziel der Buchreihe ist die Publikation neuer Ansätze in der Automation auf wissenschaftlichem Niveau, Themen, die heute und in Zukunft entscheidend sind, für die deutsche und internationale Industrie und Forschung. Initiativen wie Industrie 4.0, Industrial Internet oder Cyber-physical Systems machen dies deutlich. Die Anwendbarkeit und der industrielle Nutzen als durchgehendes Leitmotiv der Veröffentlichungen stehen dabei im Vordergrund. Durch diese Verankerung in der Praxis wird sowohl die Verständlichkeit als auch die Relevanz der Beiträge für die Industrie und für die angewandte Forschung gesichert. Diese Buchreihe möchte Lesern eine Orientierung für die neuen Technologien und deren Anwendungen geben und so zur erfolgreichen Umsetzung der Initiativen beitragen.

Weitere Bände in der Reihe http://www.springer.com/series/13886

Jürgen Beyerer · Alexander Maier ·
Oliver Niggemann
Editors

Machine Learning for Cyber Physical Systems

Selected papers from the International
Conference ML4CPS 2020

 Springer Vieweg

Editors
Jürgen Beyerer
Institut für Optronik, Systemtechnik und
Bildauswertung
Fraunhofer
Karlsruhe, Germany

Alexander Maier
MIT
Fraunhofer IOSB-INA
Lemgo, Germany

Oliver Niggemann
Helmut-Schmidt-Universität Hamburg
Hamburg, Germany

ISSN 2522-8579 ISSN 2522-8587 (electronic)
Technologien für die intelligente Automation
ISBN 978-3-662-62745-7 ISBN 978-3-662-62746-4 (eBook)
https://doi.org/10.1007/978-3-662-62746-4

This Springer Vieweg imprint is published by the registered company Springer-Verlag GmbH, DE part of Springer Nature.
The registered company address is: Heidelberger Platz 3, 14197 Berlin, Germany

Preface

Cyber Physical Systems are characterized by their ability to adapt and to learn. They analyze their environment, learn patterns, and they are able to generate predictions. Typical applications are condition monitoring, predictive maintenance, image processing and diagnosis. Machine Learning is the key technology for these developments.

The fifth conference on Machine Learning for Cyber-Physical-Systems and Industry 4.0 - ML4CPS - was held at the Fraunhofer Forum in Berlin, on March 22th - 23th 2020. The aim of the conference is to provide a forum to present new approaches, discuss experiences and to develop visions in the area of data analysis for cyber-physical systems. This book provides the proceedings of selected contributions presented at the ML4CPS 2020.

The editors would like to thank all contributors that led to a pleasant and rewarding conference. Additionally, the editors would like to thank all reviewers for sharing their time and expertise with the authors. It is hoped that these proceedings will form a valuable addition to the scientific and developmental knowledge in the research fields of machine learning, information fusion, system technologies and industry 4.0.

Prof. Dr.-Ing. Jürgen Beyerer
Dr. Alexander Maier
Prof. Dr. Oliver Niggemann

Table of Contents

Energy Profile Prediction of Milling Processes Using Machine Learning Techniques

Matthias Mühlbauer, Hubert Würschinger, Dominik Polzer and Prof. Dr.-Ing. Nico Hanenkamp

University, Erlangen-Nuremberg, Germany
Department of Mechanical Engineering
Institute of Resource and Energy Efficient Production Systems
matthias.muehlbauer@fau.de

Abstract. The prediction of the power consumption increases the transparency and the understanding of a cutting process, this delivers various potentials. Beside the planning and optimization of manufacturing processes, there are application areas in different kinds of deviation detection and condition monitoring. Due to the complicated stochastic processes during the cutting processes, analytical approaches quickly reach their limits. Since the 1980s, approaches for predicting the time or energy consumption use empirical models. Nevertheless, most of the existing models regard only static snapshots and are not able to picture the dynamic load fluctuations during the entire milling process. This paper describes a data-driven way for a more detailed prediction of the power consumption for a milling process using Machine Learning techniques. To increase the accuracy we used separate models and machine learning algorithms for different operations of the milling machine to predict the required time and energy. The merger of the individual models allows finally the accurate forecast of the load profile of the milling process for a specific machine tool. The following method introduces the whole pipeline from the data acquisition, over the preprocessing and the model building to the validation.

Keywords: energy prediction, time prediction, power load prediction, milling process, machine learning, regression

1   Einleitung

Der industrielle Sektor ist in Deutschland mit 28,9 % am Gesamtenergieverbrauch (Stand: 2016, [1]) beteiligt, wobei ein wesentlicher Teil direkt auf zerspanende Prozesse zurückzuführen ist. Zunehmender politischer Druck, strengere Regulierungen und steigende Strompreise drängen die Hersteller zu immer nachhaltigeren, energieeffizienteren Produktionsprozessen. So führt ein Ausschussteil am Ende der Prozesskette einer spanenden Produktion zu einem Energieverlust von 60 bis 80 MJ pro Kilogramm Bauteilmasse [2]. Um schon vor Produktionsbeginn den Energiebedarf eines Bearbeitungsvorgangs, beispielsweise hinsichtlich der Reduktion von Lastspitzen, optimieren zu können, muss der Produktionsprozess transparent gemacht werden. Weiter trägt eine

© The Author(s) 2021
J. Beyerer et al. (Hrsg.), *Machine Learning for Cyber Physical Systems*, Technologien für die intelligente Automation 13,
https://doi.org/10.1007/978-3-662-62746-4_1

geschaffene Transparenz zur Identifikation von Anomalien im Bearbeitungsprozess bei, welche ggf. zu Ausschuss führen können. Beides ist über die Prognose des Energiebedarfs und dem folgenden Abgleich von Soll- und Ist-Werten bzw. der Optimierung von Parametereinstellungen möglich. Analytische Verfahren zur Prognose des Energiebedarfs, Ansätze, die auf den physikalischen Gesetzen der Fertigungsverfahren beruhen, stoßen im Bereich der Zerspanung schnell an Grenzen und werden heute kaum noch eingesetzt [3]. Diese Modelle haben das grundsätzliche Problem, dass sie die komplizierten stochastischen Prozessmechanismen während der Bearbeitung nicht ausreichend abbilden [4].

Eine weitere Möglichkeit ist die Modellierung des Energiebedarfs durch empirische Ansätze. Durch den vereinfachten Datenzugriff bei automatisierten Prozessen bieten Maschinelle Lernverfahren die Möglichkeit der Erstellung dynamischer Prognosemodelle. Erste Ansätze, die solche Verfahren für die Energieprognose verwenden, wurden in den vergangenen Jahren publiziert. Kant et al. nutzt Neuronale Netze zur Vorhersage des Energiebedarfs von einfachen Fräs- [5] als auch erster Drehprozesse [6]. In weiteren Veröffentlichungen nutzt der Autor u.a. klassische Techniken des Maschinellen Lernens wie die Support Vector Regression [7]. Während sich Kant in seinen Arbeiten auf die Datenaufnahme einfacher, statischer Prozesse fokussiert, betrachtet Park [8] unterschiedliche Fertigungsarten (wie z.B. Bohren und diverse Fräsverfahren) und nutzt dafür die Gaußprozess Regression.

Park und Kant konzentrieren sich in ihren Untersuchungen nur auf die Prognose des Energiebedarfs zerspanender Bearbeitungsschritte, abgeleitet aus Numerical Control (NC)-Datensätzen. Neben diesen Prognosemethoden des Energiebedarfs wurden noch Methoden zur Prognose des Zeit- und Wegbedarfs von Zerspanungsprozessen entwickelt. Hier lassen sich unterschiedliche Herangehensweisen unter Anwendung des Maschinellen Lernens identifizieren. Saric et al. [9] entwickelte in diesem Kontext ein Modell für die Zeitprognose, mit Hilfe Neuronaler Netze. Gopalakrishnan et al. [10] erläutert die Vorhersage der Werkzeugweglänge beim Taschenfräsen durch Neuronale Netze. Diese dient wiederum als Variable für die Abschätzung der Bearbeitungszeit und der entstehenden Kosten.

Die im Rahmen dieser Veröffentlichung dargestellte Methode geht einen Schritt weiter als bisher bekannte Ansätze und erhöht den Detailgrad der Prognose so, dass eine Abbildung des gesamten Funktionsumfangs der Bearbeitungsmaschine sowie der Einzelaktionen über die Zeit ermöglicht werden. Dies erfolgt durch getrennte Modelle, die die Prognose des Zeitbedarfs und des Energiebedarfs für verschiedene Aktionsschritte der Maschine repräsentieren. Dabei lassen sich einzelne Aktionen, wie z.B. die Drehzahländerungen der Spindel oder das Verfahren der Achsen, zu eigenständigen Prognosemodellen zusammenfassen. Über die Aggregation der Einzelmodelle lässt sich eine resultierende Leistungskurve erstellen, welche Aufschluss über die anliegende Last zu jedem Zeitpunkt gibt. Durch diese neue Methode kann ein detailliertes Abbild eines Zerspanprozesses auf Basis des Energiebedarfs über die Zeit erstellt werden, wodurch sich neue Möglichkeiten der Optimierung des Energiebedarfs sowie der Erkennung von Anomalien auf Ebene einzelner Maschinenaktionen ergeben.

2   Methode

Anhand der folgenden Methode werden die Schritte von der Trainingsdatenerhebung über die Modellbildung sowie dessen Anwendung zur Prognose des Leistungsprofils beschrieben (vgl. Abb. 1). Die Vorgehensweise wird anhand einer Fräsbearbeitung aufgezeigt und validiert, jedoch ist die grundlegende Systematik auch auf weitere automatisierte Prozesse übertragbar. Der Datenvorverarbeitungs- und Modellbildungsvorgang erfolgt automatisiert und ermöglicht ein direktes Abgreifen der erforderlichen Daten während des Bearbeitungsprozesses. Die aufbereiteten Rohdaten ermöglichen im Weiteren die Erstellung von Regressionsmodellen, welche für die Prognose des Zeit- und Energiebedarfs genutzt werden. Abschließend wird das vollständige Leistungsprofil durch das Zusammensetzen aus Zeit- und Energiebedarf erstellt.

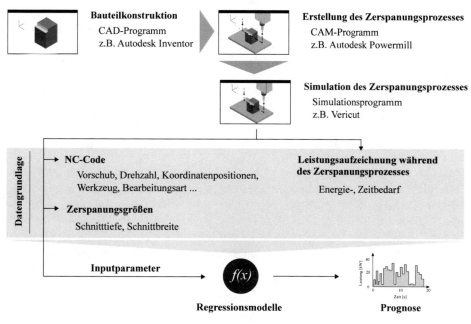

Abb. 1. Darstellung der entwickelten Methode

Der NC-Code beinhaltet bereits Informationen über Drehzahl, Vorschub, Verfahrweg und das verwendete Werkzeug. Für eine Prognose des Energiebedarfs, sowohl im Leerlauf als auch während des Zerspanvorgangs, sind weitere Daten über Schnittgrößen wie etwa Schnittiefe und Schnittbreite während der Bearbeitung erforderlich. Diese werden mithilfe einer Simulation des Zerspanvorgangs ermittelt. Die relevanten Attribute, welche für die Durchführung der Modellbildung erforderlich sind, wurden im Vorfeld definiert.

Um das Leistungsprofil in der nötigen Güte prognostizieren zu können ist es erforderlich, die Prognose des Zerspanvorgangs in einzelne Elemente bzw. Einzelaktionen aufzuteilen und dabei die Maschinencharakteristik zu einer weiteren Erhöhung der

Prognosegüte einzubeziehen. Sowohl bei Achsbewegung ohne Materialabtrag als auch bei Zerspanvorgängen selbst hat diese einen entscheidenden Einfluss auf den Zeitbedarf. Die Erstellung der Modelle auf Basis realer Datenaufzeichnungen der betrachteten Werkzeugmaschine ermöglicht eine genaue Abbildung dieser Einflussparameter.

3   Datenerhebung und -aufbereitung

Die Datenerhebung für die anschließende Modellbildung unterteilt sich in die Aufzeichnung der Zielwerte Energie- und Zeitbedarf und die Gewinnung der entsprechenden Werte der Einflussparameter. Als Datenquellen dienen hier NC-Befehlssätze, Simulation und Werkzeugmaschine. Um aufwendige Datenaufbereitungsvorgänge zu umgehen, werden für die Prognose des Leistungsverlaufs möglichst vorhandene Informationen aus den NC-Befehlssätzen genutzt. Fehlende Informationen zu Zerspanungsgrößen, wie etwa die Schnittbreite und Schnitttiefe während des Bearbeitungsvorganges, werden durch die Simulationssoftware Vericut berechnet und ausgeleitet. Dies ermöglicht eine Extraktion sämtlicher relevanter Inputparameter aus der Erstellung und Simulation des Zerspanungsvorgangs. Die Daten der Zielvariablen Energie- und Zeitbedarf werden durch Versuche an der Maschine aufgezeichnet und durch Schnittstellen extrahiert.

3.1   Gewinnung der Zielwerte Energie- und Zeitbedarf

Das im Rahmen dieser Arbeit betrachtete Bearbeitungszentrum (Doosan DNM 500 II) ermöglicht die Aufzeichnung der Leistungsaufnahme der einzelnen Verbraucher über eine Siemens Numerik Steuerung (Sinumerik 828D). Diese Leistungsaufzeichnungen dienen als Grundlage für die nachfolgende Modellbildung. Um die erforderliche Datenbasis für die Modellbildung zu schaffen, wurden bei insgesamt 32 Planfräsprozessen, 186 Verfahrvorgänge der Achsen und 30 Drehzahländerungsvorgänge die Wirkleistungsverläufe der wesentlichen Verbraucher, der drei Achsen und der Spindel, aufgezeichnet. Es wurde eine Routine entwickelt, um die Daten automatisiert aufzubereiten und vorzuverarbeiten. Während der Aufbereitung der Daten werden die Energie- und Zeitbedarfe der einzelnen Bearbeitungsschritte berechnet und die Werte der Eingangs- und Zielvariablen in den entsprechenden Datensätzen für das Anlernen der einzelnen Modelle abgespeichert.

3.2   Gewinnung der Inputparameter für die Regressionsmodelle

Um den Energiebedarf auch bei einem Werkzeugeingriff ausreichend genau abbilden zu können, sind neben den Daten aus den NC-Befehlen Informationen zu den Spanungsgrößen erforderlich. Für die Gewinnung der für die Prognose relevanten Daten wird die Simulationssoftware Vericut verwendet. Diese ermöglicht eine Simulation der Werkzeugwege und der Materialabtragsprozesse. Die hierbei berechneten Spanungsgrößen können zusammen mit den NC-Datensätzen aus dem Programm ausgeleitet werden und bilden die Basis für die Erstellung der Regressionsmodelle. Für die anschließende Prognose des Leistungsverlaufs eines betrachteten Zerspanungsprozesses wird ebenso verfahren. Die Zerspanungsprozesse werden in der Regel mit Hilfe einer CAM-Software simuliert. Eine zusätzliche Überführung des Bearbeitungsprogrammes in Vericut liefert weitere Inputparameter, welche für die Energie- und Zeitprognose erforderlich sind.

3.3 Feature Engineering

Im Falle der Energieprognose gilt es, über Feature Engineering, die Einflussparameter zu identifizieren, die Auswirkung auf den Energiebedarf haben und in ihrer Gesamtheit somit eine möglichst genaue Prognose ermöglichen. Anhaltspunkte hierfür geben die Literatur, Domänenwissen und die bereits bestehenden Modelle zur Berechnung der Schnittkraft. Kühn et al. [11] nennt zehn Einflussgrößen, die sich auf die Schnittkraft F_C und somit direkt auf die benötigte Leistung auswirken: Werkstoff, Vorschub bzw. Spanungsdicke, Schnitttiefe bzw. Spanungsbreite, Spanungsverhältnis (Schnitttiefe/Vorschub), Spanwinkel, Einstellwinkel, Schnittgeschwindigkeit, Schneidstoff, Kühlung und Schmierung und Werkzeugverschleiß.

4 Modellbildung

Da mit dieser Methode der Energiebedarf des gesamten Funktionsumfanges der Bearbeitungsmaschine, repräsentiert durch unterschiedliche Verbraucher, abgedeckt werden soll und sich die Einflussparameter auf den Energiebedarf je nach Aktionsschritt stark unterscheiden, erfolgt die Modellbildung auf Einzelaktionsebene, siehe Abb. 2.

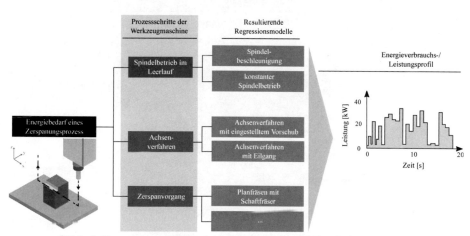

Abb. 2. Regressionsmodelle der Einzelaktionen des Bearbeitungsprozesses

Zum einen werden verschiedene Modelle für zeitgleich aktive Verbraucher, wie z. B. Spindelantrieb und Achsantriebe, erstellt. Zum anderen werden auch für dieselben Verbraucher verschiedene Modelle erstellt, wenn diese stark unterschiedliche Aktionen durchführen, wie z. B. das Verfahren der Achsen im Normalbetrieb bzw. im Eilgang.

Die Aufteilung in verschiedene Einzelaktionen bzw. Modelle erfolgte mit der Prämisse der Maximierung der Leistungsfähigkeit der Einzelmodelle und somit der Prognosegenauigkeit unter der Berücksichtigung einer noch vertretbaren Anzahl sich ergebender Modelle. Die erstellten Einzelmodelle werden abschließend in einem Gesamtmodell aggregiert, welches die Prognose des Leistungsprofils ermöglicht.

Konnte eine definierte Einzelaktion nicht ausreichend durch ein Modell abgebildet werden, erfolgte eine detailliertere Untersuchung der zugrundeliegenden Datenstruktur. Beispielsweise wurde beim Zeitbedarf für das Verfahren der Achsen im Eilgang eine Knickstelle bei der Analyse der Daten festgestellt. Durch diese spezielle Datenstruktur ist eine separate Modellbildung links und rechts der Knickstelle sinnvoll, siehe Abb. 3. Weder die Gaußprozess Regression noch die Polynomiale Regression können die vollständigen Daten richtig annähern. Durch die Aufteilung in zwei Modelle gelingt eine verbesserte Annäherung.

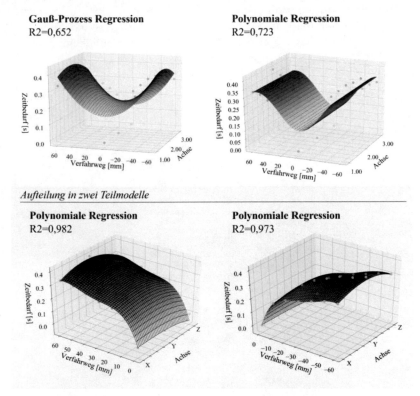

Abb. 3. Verbesserung der Prognosegüte durch Modellierung von zwei Teilmodellen

Während des Modellbildungsprozesses wurden für die einzelnen Datensätze der definierten Aktionen unterschiedliche Regressionsverfahren mit Hilfe der Kreuzvalidierung (Leave-One-Out) evaluiert. Betrachtet wurde dabei die Leistungsfähigkeit der Linearen, Polynomialen, Gaußprozess und Random Forest Regression über die Evaluationsmetrik Mittlere Quadratische Abweichung (MSE) und die Mittlere Absolute Abweichung (MAE). Über eine Routine konnten die Hyperparameter der Verfahren automatisch variiert und bewertet werden. Das Gütemaß MSE diente hierbei als Optimierungsgröße. In Tabelle 1 werden die Unterschiede der Regressionsverfahren am Beispiel der Aktion Drehzahländerung verdeutlicht. Je Einzelaktion wurde das Regressionsverfahren bestimmt, welches die geringste Abweichung (MSE) aufweist und somit die jeweilige Aktion am besten annähert. Das finale Gesamtmodell zur Prognose des Leistungsverlaufs setzt sich dementsprechend aus den elf gewählten Teilmodellen zusammen.

Tab. 1. Vergleich der Regressionsverfahren zur Prognose des Energiebedarfs der Aktion Drehzahländerung

Regressionsverfahren	Hyperparameter	MAE [kJ]	MSE [kJ²]
Lineare Regression		3,21	13,30
Polynomiale Regression	Polynomgrad=2	0,55	0,69
Random Forest	Bäume=10; max. Tiefe=10	2,62	9,29
Gaußprozess Regression	Kernel = RBF	0,81	1,05

5   Ergebnisse und Validierung

Sämtliche Regressionsmodelle sind speziell auf die betrachtete Bearbeitungsmaschine abgestimmt und bilden somit Maschinenspezifika mit ab. Dies ermöglicht hohe Prognosegenauigkeiten.

Tab. 2. Teilregressionsmodelle für den Energiebedarf von Zerspanungsprozessen

Regressionsverfahren	Hyperparameter	Attribute	R^2	MSE [kJ²]	
Energiemodell: Achsenverfahren im Eilgang					
Polynomiale Regression	Polynomgrad=4	Verfahrweg (und Richtung) [mm], Achse [x=1, y=2, z=3]	0,986	2,07 10^{-4}	*
Energiemodell: Achsenverfahren mit eingestelltem Vorschub					
Polynomiale Regression	Polynomgrad=4	Verfahrweg (und Richtung) [mm], Achse [x=1, y=2, z=3], Vorschub f [mm/min]	0,982	1,72 10^{-4}	*
Energiemodell: Drehzahländerung					
Polynomiale Regression	Polynomgrad=2	Ausgangsdrehzahl [1/min], Enddrehzahl [1/min]	0,992	0,26	
Energiemodell: Werkzeug (Schaftfräser) im Eingriff					
Polynomiale Regression	Polynomgrad=2	Verfahrweg [mm], Vorschub [mm/min], Drehzahl [1/min], Schnitttiefe [mm], Schnittbreite [mm]	0,980	0,22	
Energiemodell: konstanter Spindelbetrieb					
Random Forest	Bäume=10; Max. Tiefe=20	Drehzahl n [1/min]	0,604	0,11	

Die weitere Evaluation der Modelle erfolgte auf Basis eines Testdatensatzes. Tabelle 2 zeigt die für die Energieprognose verwendeten Regressionsverfahren, die relevanten Hyperparametereinstellungen, die Merkmale der Modelle und die Evaluationsergebnisse anhand des Bestimmtheitsmaßes (R^2) und der mittleren quadratischen Abweichung (MSE).

Das Bestimmtheitsmaß nimmt bei den meisten Modellen einen Wert über 0,9 ein. Die Ausnahme bildet hier das Modell für die Prognose des Energiebedarfs bei konstantem Spindelbetrieb. Es wird angenommen, dass der verwendete Datensatz mit lediglich sieben Instanzen ursächlich für die geringere Modellgüte ist. Die Ergebnisse der mittleren quadratischen Abweichungen (MSE) der Prognosewerte spiegeln diese Ergebnisse wieder.

Hiermit konnte gezeigt werden, dass eine Prognose des Energiebedarfs auf Einzelaktionsebene durch die gewählten Verfahren und Attribute mit hoher Genauigkeit möglich ist. Gleichwertige Ergebnisse konnten für die Prognose des Zeitbedarfs erzielt werden (vgl. Tabelle 3). Die Modelle für den Zeitbedarf erreichen durchgehend ein Bestimmtheitsmaß über 0,9.

Tab. 3. Teilregressionsmodelle für den Zeitbedarf von Zerspanungsprozessen

Regressionsverfahren	Hyperparameter	Attribute	R^2	MSE [s^2]
Zeitmodell: Achsenverfahren im Eilgang (in positive Koordinatenrichtung)				
Polynomiale Regression	Polynomgrad=4	Verfahrweg (und Richtung) [mm], Achse [x=1, y=2, z=3]	0,935	1,16 * 10^{-3}
Zeitmodell: Achsenverfahren im Eilgang (in negative Koordinatenrichtung)				
Polynomiale Regression	Polynomgrad=4	Verfahrweg (und Richtung) [mm], Achse [x=1, y=2, z=3]	0,982	3,19 * 10^{-4}
Zeitmodell: Achsenverfahren mit eingestelltem Vorschub				
Polynomiale Regression	Bäume=20; Max. Tiefe=20	Verfahrweg (und Richtung) [mm], Achse [x=1, y=2, z=3], Vorschub f [mm/min]	0,994	1,85 * 10^{-2}
Zeitmodell: Drehzahländerung				
Gauß-Prozess Regression	Rational Quadratic Kernel; Alpha=0,1; Varianz=1,0; Lenghtscale=1,0	Ausgangsdrehzahl [1/min], Enddrehzahl [1/min]	0,942	6,37 * 10^{-3}

Die Prognosewerte werden im Weiteren zur Berechnung und Darstellung des gesamten Leistungsverlaufs genutzt. Auf Basis der prognostizierten Zeitbedarfe für die Einzelaktionen und der Berücksichtigung, ob diese parallel bzw. seriell stattfinden, ergibt sich die geplante Dauer der Einzelaktion und des gesamten Bearbeitungsvorgangs. Die prognostizierten Energiebedarfe werden den entsprechenden Zeitbedarfen zugeordnet, wodurch der Leistungsverlauf ausgeprägt wird und folgend die Ermittlung des Gesamtenergiebedarfs durch Integration ermöglicht. Abb. 4 stellt die reale Leistungsaufzeichnung während eines Planfräsprozesses und die der Prognose gegenüber. Die prognostizierte Leistungskurve spiegelt den realen Verlauf erkennbar wider. Weiter decken sich die prognostizierten und realen Energiebedarfswerte für die definierten Aktionen. Die absolute Abweichung (MAE) des Beispielprozesses eines Planfräsvorgangs beträgt 1,02 kJ, was einer Abweichung von unter 4 % entspricht.

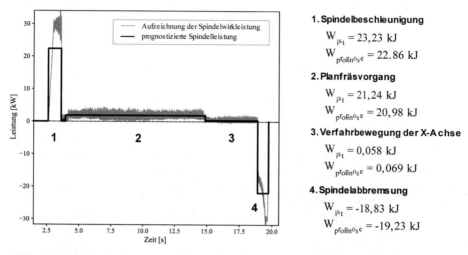

1. **Spindelbeschleunigung**

$W_{ist} = 23{,}23$ kJ

$W_{prognose} = 22.86$ kJ

2. **Planfräsvorgang**

$W_{ist} = 21{,}24$ kJ

$W_{prognose} = 20{,}98$ kJ

3. **Verfahrbewegung der X-Achse**

$W_{ist} = 0{,}058$ kJ

$W_{prognose} = 0{,}069$ kJ

4. **Spindelabbremsung**

$W_{ist} = -18{,}83$ kJ

$W_{prognose} = -19{,}23$ kJ

Abb. 4. Vergleich der Prognose und des aufgezeichneten Leistungsverlaufs während eines Zerspanungsprozesses (Beispiel: Planfräsvorgang mit einem 12 mm Schaftfräser, Drehzahl: 7000 1/min, Vorschub: 560 mm/min, Schnitttiefe: 6 mm, Schnittbreite: 9 mm)

6   Diskussion und Ausblick

Die vorgestellte Methode demonstriert die Prognose des Leistungsprofils bzw. des Energiebedarfs eines Bearbeitungszentrums mit hoher Genauigkeit auf Einzelaktionsebene. Hierdurch ergeben sich Potentiale im Hinblick auf diverse Optimierungsmaßnahmen. Diese ergeben sich durch die Optimierung von Prozessparametern hinsichtlich der Minimierung des Zeit- oder Energiebedarfs des Zerspanvorgangs, dem Mikro-Lastmanagement bzw. der Vermeidung von Lastspitzen bei der Berücksichtigung von mehreren parallellaufenden Prozessen und auch im Bereich der Anomaliedetektion durch einen Vergleich von Soll- und Ist-Werten.

Obwohl die beschriebene Methode schon weitestgehend automatisiert wurde, ist eine weitere Automatisierung der Vorgehensweise anzustreben. Insbesondere die Erhebung von maschinenspezifischen Trainingsdaten gestaltet sich noch aufwändig. Hier ist ein einfach implementierbares Edge-Device mit den erforderlichen Schnittstellen zur Datenerhebung an der Maschine und mit der Möglichkeit zur Datenvorverarbeitung denkbar. Um die genannten Potentiale zur Anomalieerkennung ausschöpfen zu können, ist zudem ein Abgleich von Soll- und Ist-Werten in Echtzeit erforderlich.

References

1. Umweltbundesamt auf Basis Arbeitsgemeinschaft Energiebilanzen: Endenergieverbrauch 2017 nach Sektoren. Auswertungstabellen zur Energiebilanz für die Bundesrepublik Deutschland 1990 bis 2017, Stand 07/2018
2. Fraunhofer Gesellschaft, Energieeffizienz in der Produktion: Untersuchung zum Handlungs- und Forschungsbedarf. München (2008)
3. Bhinge, R.; Park, J.; Law, K. H.; Dornfeld, D. A.; Helu, M.; Rachuri, S.: Toward a Generalized Energy Prediction Model of Machine Tools. In: Journal of Manufacturing Science and Engineering (2017)
4. Duerden, C.; Shark, L.-K.; Hall, G; Howe, J.: Prediction of Granular Time-Series Energy Consumption for Manufacturing Jobs from Analysis and Learning of Historical Data. In: 2016 Annual Conference on Information Science and Systems (2016)
5. Kant, G.; Sangwan, K. S.: Predictive Modelling for Energy Consumption in Machining Using Artificial Neural Network. In: Procedia CIRP (2015)
6. Sangwan, K. S.; Kant, G.: Optimization of Machining Parameters for Improving Energy Efficiency using Integrated Response Surface Methodology and Genetic Algorithm Approach. In: Procedia CIRP (2017)
7. Kant, G.: Prediction and Optimization of Machining Parameters for Minimizing Surface Roughness and Power Consumption during Turning of AISI 1045 Steel. Birla Institute of Technology & Science, Pilani (2016)
8. Park, J.; Law, K. H.; Bhinge, R.; Biswas, N.: A Generalized Data-Driven Energy Prediction Model with Uncertainty for a Milling Machine Tool Using Gaussian Process. In: Volume 2: Materials; Biomanufacturing; Properties, Applications and Systems; Sustainable Manufacturing (2015)
9. Saric, T.; Simunovic, G.; Simunovic, K.: Estimation of Machining Time for CNC Manufacturing Using Neural Computing. In: International Journal of Simulation Modelling (2016)
10. Gopalakrishnan, B.; Reddy, V. K.; Gupta, D. P.: Neural networks for estimating the tool path length in concurrent engineering applications. In: Journal of Intelligent Manufacturing (2004)
11. Kühn, K.-D.; Fritz, A. H.; Förster, R.; Hoffmeister, H.-W.: Trennen. In: Fertigungstechnik. 12. Auflage, Hrsg. Fritz, A. H., Springer Vieweg, Berlin (2018)

Improvement of the prediction quality of electrical load profiles with artificial neural networks

Fabian Bauer[1], Jessica Hagner[1], Peter Bretschneider[1,2], Stefan Klaiber[2]

[1] Technische Universität Ilmenau, Energy Usage Optimization Group, Gustav-Kirchhoff-Str. 5, 98693 Ilmenau, Germany
{fabian.bauer,jessica.hagner,peter.bretschneider}@tu-ilmenau.de
[2] Fraunhofer Advanced System Technology (AST) Branch of Fraunhofer IOSB, Am Vogelherd 50, 98693 Ilmenau, Germany
stefan.klaiber@isob-ast.fraunhofer.de

Abstract. Against the backdrop of the economically and ecologically optimal management of electrical energy systems, accurate predictions of consumption load profiles play an important role. On this basis, it is possible to plan and implement the use of controllable energy generation and storage systems as well as energy procurement with the required lead-time, taking into account the technical and contractual boundary conditions.

The recorded electrical load profiles will increase considerably in the course of the digitization of the energy industry. In order to make the most accurate predictions possible, it is necessary to develop and investigate models that take account of the growing quantity structure and, due to the significantly higher number of observations, improve the forecasting quality as far as possible.

Artificial neural networks (ANN) are increasingly being used to solve non-linear problems for a growing amount of data that is affected by human and other unpredictable influences. Consequently, the model approach of an ANN is chosen for predicting load profiles. Aim of the thesis is the simulative investigation and the evaluation of the quality and optimality of a prediction model based on an ANN for electrical load profiles.

Keywords: Artificial Neural Network, Electrical Load Prediction, Machine Learning.

1 Introduction

Accurate forecasts of future events are crucial for the optimal operation of electrical energy systems. Due to the digitization of the energy industry, the number of recorded electrical load profiles and also the number of observations per time series will increase significantly. In order to analyze the growing amount of data, corresponding systems and models are needed that are able to recognize structures and patterns from complex information.

© The Author(s) 2021
J. Beyerer et al. (Hrsg.), *Machine Learning for Cyber Physical Systems*, Technologien für die intelligente Automation 13,
https://doi.org/10.1007/978-3-662-62746-4_2

Machine learning methods offer an efficient alternative to the manual extraction of the knowledge contained in the data and the derivation of rules. They are used to extracting correlations and insights from large amounts of data in order to make predictions about future events. [1,2]

ANN are among the nonlinear dynamic models. The input and output behavior is represented by observations of the process, whereby the connections are represented by internal structures. In recent years, advances in research and technology, especially in computing power and algorithms, have led to better procedures through neural networks. By processing large amounts of data and nonlinear relationships, ANN are very successful, for example in speech recognition. [3-5]

2   Analysis of the load profiles

The power consumption data is available in the form of annual load profiles. The load curve shows the time course of the electrical power drawn. The time resolution is 15 minutes. Accordingly, 4 measured values are available per hour and 96 measured values on one day. Consequently, the annual load curve is mapped with 35040 data points.

2.1   Primary data preparation and plausibility check

The first step before modelling is the preparation of the primary data. The time series are checked for possible measurement errors, outliers and for completeness. In the case of individual missing measured values or obvious measurement errors, these substitute values are interpolated. Due to the normalization of the time series, the values of the load profiles are between zero and one. In the course of the plausibility check, any zero values that occur are set from zero to a low value due to the normalization in the load profile. Otherwise, this data would not represent any relevant information in the neural network and would be weighted with zero. An actual value of zero can only be expected in the event of a power failure, which is not relevant for the load curve forecast.

2.2   Data analysis and creation load profile classes

To study the model approach, five load profiles are used, which were examined and classified for selected statistical parameters and show a typical load curve for households. This can be seen from the morning and evening peaks during the day and the weekly rhythm.

2.3   Parameter estimation

After defining the model approach, the network parameters are estimated in the course of modeling. The estimation of the parameters, i.e. the edge weights and threshold values, is achieved by training the net. In order to be able to train the network in the best possible way, an appropriate input assignment is coordinated and adapted. Furthermore, data sets are divided into training and test data.

The selection of the additional historical values was determined using the autocorrelation coefficient (ACF) and the partial autocorrelation coefficient (PACF) with regard to the relevant information content for the forecast. The aim is to reduce the complexity of the ANN in terms of computing power and time.

Table 1 lists six historical depths that are used as input variables for the model.

Table 1. Autocorrelation coefficients for the historical values

Historical values	relation	ACF	PACF
x(k-1)	quarter hour before	0,9957	0,9957
x(k-2)	half hour before	0,9855	-0,6986
x(k-3)		0,9698	-0,2954
x(k-96)	day before	0,8391	-0,2386
x(k-97)		0,8357	-0,2330
x(k-673)	week before	0,9365	-0,2252

2.4   Splitting the data sets

The time series are divided into two subsets for the analysis of the ANN. The time series were divided 2/3 to 1/3 into training and test data. Therefore the neural network is trained with eight months of the time series. The remaining four months of the year (11615 measured values) are forecasted using the trained network, compared with the test data and evaluated.

3   Artificial neural network as prediction model

An artificial neural network is an information processing system with analogies to the human brain. The basic idea is based on the reproduction of biological nerve cells and the associated neuronal connections, which can be used to reproduce complex processes. The advantage of ANN is their ability to learn. They are able to adapt to changing conditions and to learn further on the basis of additional data. This makes it possible to continuously improve the network and increase the forecasting quality. [6]

The ANN for the prediction model is constructed as a feedforward network due to abstraction and is illustrated in **Fig. 1**. There are various state-of-the-art ANN methods for similar tasks, such as recurrent neural networks, that exhibit time-dynamic behavior through an internal memory.

The information is transmitted between the nodes, via the connection. The connection weighting varies the data transfer and passes it on to the next layer.

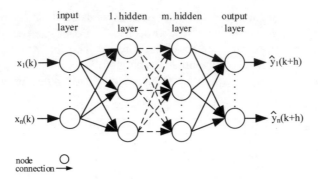

Fig. 1. Exemplary ANN set up as feedforward network for electric load profiles prediction

Between the input and output layer there can be a number of hidden layers. The system learns by changing the connection weights. The sum of the inputs is weighted in the nodes and transferred to the transfer function when a threshold value is exceeded. The result serves as input for the next layer.

The neural network structure is trained using a backpropagation algorithm. For that the net weights and threshold values are corrected by means of the gradient descent procedure to minimize errors. The prediction of electrical load profiles is carried out by the trained ANN by applying n historical values in the input layer. Hereupon the trained net is used to map the forecast value in the output layer. This model approach serves as starting point for the following investigations.

3.1   Research studies

The investigation of the prediction model based on an ANN to forecast electrical load profiles are the subject of a current research project at the department. In that respect the model structure of the ANN is mapped and simulated in Matlab and Python. The present paper examines different network structures and hyperparameters to find the optimal network configuration to forecast electrical load profiles. The investigation should show if and what advantage (deep) neural networks have in forecasting of electrical load profiles.

The focal points of the investigation related to the prediction quality are:
- Influence of the number of network nodes
- Training of the ANN with unstructured/ structured Data
- Influence of the data quantity
- Selection of historical values/depth of historical values

The research focuses on how large amounts of data can be integrated into the model. The investigations aim to the influence of the data quantity related with variation of the network structure and node number and the stability of the forecast. As a result, statements can be made about quality and optimality of the model approach.

3.2   Basic specifications of the model

As input variable x(t) a sliding time window with n historical values is applied over the electrical annual load curve. **Fig. 2** illustrates the electrical load profile of a household customer as input for the prediction model.

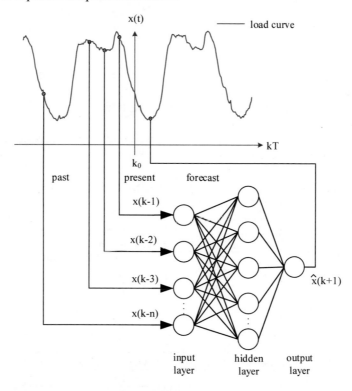

Fig. 2. Example of the prediction model for electrical load profiles form n historical data

The chosen model approach of the ANN has the following basic structure:
- 6 input nodes
- 1 hidden layer
- 10 knots in the hidden layer
- 1 output node
- Sigmoid activation function
- Fully meshed network

Starting from this network, the parameter estimates and the simulative investigation are carried out. In order to obtain precise conclusions on the mode of operation and the change in the quality of the forecast, the structure of the network is modified in further investigation steps.

3.3   Investigation scenarios

In the training phase, the neural network learns to deduce the output variables from the given input variables. The forecast model is trained on the prediction horizon x(k+96). During the training the connections of the nodes are weighted and determined. The training will continue until the performance goal is reached. Derived from the main points of investigation in Chapter 3.1, the structure of the network is modified on the basis of the basic model.

The following changes made for the analysis of the model behavior:
- Increase of the data quantity with further time series
- Increase of the number of network nodes in the hidden layer
- Increase of the output assignment

For each change in the structure of the neural network, a scenario is created (**Table 2**). The individual changes are also combined with each other.

Table 2. Investigation scenarios with the respective network configuration

Scenario	Timeseries	Notes		
		Input	Hidden Layer	Output
Variant 1: Simultaneous connection of time series				
1	1	6	10	1
2	1	6	50	1
3	5	30	10	1
4	5	30	10	5
5	5	30	50	1
6	5	30	50	5
Variant 2: Continuous connection of the time series				
7	5	6	10	1
8	5	6	50	1

4   Simulation and evaluation of the results

After training the ANN, the simulation is based on the test data that was not used for the training. The forecast generated by the model is then compared with the recorded measured values.

For the validation and evaluation of the results, the residuals, i.e. the error between prognosis and observations, are used. The average forecast error and the mean absolute error (mae) can be calculated on this basis.

$$\text{mae} = \frac{1}{N} \cdot \sum_{k=1}^{N} |\hat{x}_k - x_k| \tag{1}$$

The advantage of the mae over the mean error is the investigation of the absolute deviations between forecast and observation.

The quality of the forecast is also evaluated from the error measures and makes a statement about how well the forecast value corresponds to the real value.

$$\text{quality} = \frac{1}{N} \cdot \sum_{k=1}^{N}\left(1 - \frac{|\hat{x}_k - x_k|}{x_k}\right) \tag{2}$$

In addition, r is calculated, resulting in the linear relationship between forecast and observation.

$$r = \frac{\frac{1}{N}\sum_{k=1}^{N}(x_k - \overline{x_k})(\hat{x}_k - \overline{\hat{x}_k})}{\sqrt{\frac{1}{N}\sum_{k=1}^{N}(x_k - \overline{x_k})^2} \cdot \sqrt{\frac{1}{N}\sum_{k=1}^{N}(\hat{x}_k - \overline{\hat{x}_k})^2}} \tag{3}$$

Table 3 compares the results for the 8 scenarios examined (cf. **Table 2**) for the forecast of the value with the prediction horizon x(k+96).

Table 3. Results by scenarios for prediction horizon x(k+96)

Scenario	$\overline{f}_{pred}$	mae	quality	r
1	-0,0156	0,0733	0,8596	0,8926
2	-0,0142	0,0755	0,8554	0,8887
3	-0,0400	0,0623	0,8742	0,9279
4	-0,0279	0,0595	0,8896	0,9507
5	-0,0326	0,0625	0,8737	0,9243
6	-0,0290	0,0557	0,8899	0,9415
7	0,0030	0,0719	0,8672	0,8897
8	0,0015	0,0719	0,8678	0,8916

Due to the increase in input occupancy by adding further time series from the corresponding time series class, the quality of the forecast could be improved. Furthermore, an increase in the output occupancy, with a simultaneous increase in the number of network nodes, leads to a further increase in forecast quality. This relationship is also expressed in a reduction of the mean absolute error and the increase of the correlation coefficient.

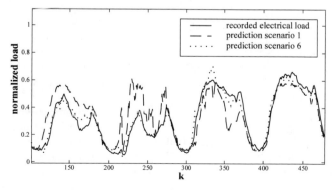

Fig. 3. Forecast load profile for scenario 1 and scenario 6 for prediction horizon x(k+96)

The time series of the forecasts for scenarios 1 and 6 are compared with the real data in **Fig. 3**. The improvement in prediction quality is the result of the structural changes in the neural network. Based on a larger amount of data and as a result a more complex network structure, the ANN can better deduce the required output variables from the given input variables during training.

5   Conclusion and Outlook

The predictive quality of electrical load profiles was examined using an ANN. As an example, the model was designed for the prediction horizon $x(k+96)$. After the various scenarios have been taught and tested, an improvement in quality can be observed with increasing data volume, accompanied by a smaller error. The increase of the number of network nodes in the hidden layer, as well as the output occupancy have a supporting effect. The variation and increase of the parameters are limited with regard to the complexity and computing power of the ANN. Another aspect of the study is the consideration of the complete period for the next 24 h in a model, based on the representative consideration of the value for the time after 24 h

Furthermore, the results form the starting point for further investigations with regard to deep learning procedures. These represent ANNs with an extensive network structure, which are able to continue learning from incoming data while application is running. [1-3]

In addition, the comparison of the forecast results by ANN with state-space models is recommended. For example, exponential smoothing and autoregressive moving average models can be considered.

References

1. Raschka, S.: Python Machine Learning. Mitp Verlag, Frechen (2017).
2. Rashid, T.: Neuronale Netze selbst programmieren. dpunkt Verlag, Heidelberg (2017).
3. Hinton, G., Deng, L., Yu, D., Dahl, G.E., Mohamed, A., Jaitly, N., Senior, A., Vanhouke, V., Nguyen, P., Sainath, T.N., Kingsbury, B.: Deep Neural Networks for Acoustic Modeling in Speech Recognition. IEEE Signal Processing Magazine (Volume: 29, Issue: 6, Nov 2012). doi:10.1109/MSP.2012.2205597.
4. AlFuhaid, A.S., El-Sayed, M.A., Mahmoud, M.S.: Cascaded artificial neural networks for short-term load forecasting. IEEE Transactions on Power Systems (Volume: 12, Issue: 4, Nov 1997). doi:10.1109/59.627852.
5. Senjyu, T., Mandal, P., Uezato, K., Funabashi, T.: Next day load curve forecasting using recurrent neural network structure. IEE Proceedings – Generation, Transmission and Distribution (Volume: 151, Issue: 3, May 2004). doi:10.1049/ip-gtd:20040356.
6. Thiesing F.: Vorhersagen mit Neuronalen Netzen. Available: http://www2.inf.uos.de/papers_html/um_5_96/start.html, last accessed (20.03.2018)

Detection and localization of an underwater docking station in acoustic images using machine learning and generalized fuzzy hough transform

Divas Karimanzira and Helge Renkewitz

Department of Surface and Maritime systems, Fraunhofer IOSB-AST, Ilmenau, Am Vogelherd 50, 98689, Ilmenau, Germany
divas.karimanzira@iosb-ast.fraunhofer.de
http://www.iosb-ast.fraunhofer.de

Abstract. Long underwater operations with autonomous battery charging and data transmission require an Autonomous Underwater Vehicle (AUV) with docking capability, which in turn presume the detection and localization of the docking station. Object detection and localization in sonar images is a very difficult task due to acoustic image problems such as, non-homogeneous resolution, non-uniform intensity, speckle noise, acoustic shadowing, acoustic reverberation and multipath problems. As for detection methods which are invariant to rotations, scale and shifts, the Generalized Fuzzy Hough Transform (GFHT) has proven to be a very powerful tool for arbitrary template detection in a noisy, blurred or even a distorted image, but it is associated with a practical drawback in computation time due to sliding window approach, especially if rotation and scaling invariance is taken into account. In this paper we use the fact that the docking station is made out of aluminum profiles which can easily be isolated using segmentation and classified by a Support Vector Machine (SVM) to enable selective search for the GFHT. After identification of the profile locations, GFHT is applied selectively at these locations for template matching producing the heading and position of the docking station. Further, this paper describes in detail the experiments that validate the methodology.

Keywords: Object detection and localization · Classifier · Generalized fuzzy Hough Transform · Selective search.

1 Introduction

Extended underwater applications such as the inspection and maintenance of underwater structures require an autonomous underwater vehicle(AUV) with autonomous docking capability for battery charging and data transmission. Underwater docking is a complex process and is composed of several sub-tasks, i.e., detection and localization, homing, physically attach to recharge AUV batteries and to establish a communication link, to wait in a low power state for a new mission, and to undock. In this paper we will focus on the *detection and localization* part of the docking process. It is assumed that the AUV is equipped with a

J. Beyerer et al. (Hrsg.), *Machine Learning for Cyber Physical Systems*, Technologien für die intelligente Automation 13,
https://doi.org/10.1007/978-3-662-62746-4_3

forward looking imaging sonar (FLS) as the perception system. Although sonars are not limited by turbidity, their data have some characteristics that make it difficult to process and extract valuable information. These characteristics are given in [1], and include non-homogeneous resolution, non-uniform intensity, speckle noise, acoustic shadowing and reverberation and multipath problem. In addition, the data are often only cross sections of the objects. In literature, some works which propose strategies to identify objects in acoustic images can be found, e.g. [1, 2]. Santos et al., developed a special system which uses acoustic images acquired by FLS to create a semantic map of a scene [1].

Acoustic images require preprocessing. Therefore, many works have been carried out for filtering and enhancement of acoustic images. Very important is the insonification described in [3, 4] where a sonar insonification pattern (SIP) obtained from averaging a large number of acoustic images taken from the same position is applied to each acoustic image reducing the effects of the non-uniform insonification and the overlapping problem of acoustic beams. Other artifacts such as speckle noise can thereafter be removed by Lee filtering [7]. Image enhancement intensifies the features of images. In [3], a method specifically developed for enhancing underwater images known as mixed Contrast Limited Adaptive Histogram Equalization (CLAHE) was discussed. Its results show less mean square error and high peak signal to noise ratio (PSNR) than other methods. Another technique for enhancement of acoustic images is using Dynamic Stochastic Resonance (DSR) technique. It has been used for enhancement of dark low contrast images in [5].

In image classification tasks, having proposal of the locations of potential objects tends to increase classification accuracy. Modern methods are based on a way to selectively search images for potential object locations by using a variety of segmentation techniques to come up with potential object hypotheses. For example, thresholding technique is used to group pixels in which input gray scale image is converted into binary image based on some threshold value [6]. Machado et al. proposed a method specifically for acoustic images [4], where the regions of interest are extracted by a linear search finding pixels with intensities higher than a certain value.

As for the detection methods, which are invariant to rotations, shifts and scale changes of objects, the Generalized Hough Transform (GHT), a geometric hashing, and variations on these methods have been proposed so far. However, the major weakness of GHT is that the scale and rotation of the object is handled in a brute-force approach which requires a four-dimensional parameter space and high computational cost.

In this paper, a method is provided which detect and recognize a docking station in a scene. The captured acoustic image is segmented, and the shape of each segment is described geometrically. Each shape is then classified into two main classes (aluprofile, and obstacles) using the well-known Support Vector Machine (SVM) algorithm. After identification of the aluprofile locations, GFHT is applied at these locations for template matching producing the heading and position of the docking station.

2   Methodology

The task is to find the location of the docking station in an acoustic image which is rather noisy by optical imaging standards. Therefore, the proposed method illustrated in Fig. 1 has six steps which include data collection, image filtering and enhancement, segmentation, segment description, classification and localization. In the first step, four different filters, each of which aims to revise a special defect

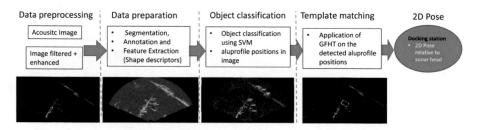

Fig. 1. Proposed methodology pipeline

are applied to the acoustic images. In the second step an automatic segmentation process of the images based on intensity peak analysis is conducted. In the third step the segments are converted to Gaussians that are easily described by shape descriptors such as the width, height, inertia ratio, area, hull area, convexity and pixel intensities information. The shape descriptors form the feature vectors which are applied in the pre-last step where a Support Vector Machine is trained to recognize the aluminum profiles of the docking station and finally GFHT [8] is applied to localize the complete docking station. In the following sections the methods applied in all steps will be discussed.

Step 1: Image denoising and enhancement . The acoustic image runs through a filtering pipeline to mitigate sonar defects starting from the non-uniform intensity problem through speckle removal and finally image enhancement using dynamic stochastic resonance. The first processing stage is about blurring the homogeneous regions keeping edges unharmed. Therefore, in this step we apply an image correction process to mitigate the non-uniform intensity problem and speckle noise. Typically this problem can be reduced by a mechanism to compensate the signal loss according to the distance traveled. However, the intensity variations can also have other causes, e.g., by changing the sonar tilt angle. As in [3, 4], we first compute the sonar insonification pattern by averaging a significant number of images captured by the sonar at the same spot. The insonification pattern is applied to each acoustic image reducing the defects. Now, having a pattern-free image, the remaining speckle noise and acoustic reverberation and multipath problem can be reduced in the next steps. Several filters for eliminating speckle based upon diverse mathematical models of the phenomenon exist. Speckle elimination in wavelet domain is very popular, but has some drawbacks, e.g., the selection of appropriate threshold is very difficult. Another method, the adaptive speckle filtering include the Lee filtering technique [7] which is based on minimum mean square error with preserving edges and the Lee filter has a special property that it converts the multiplicative model into an additive one, thereby reducing the problem of dealing with speckle noise to a known tractable case. Principally, the Lee filter works the same as the Kalman filter. During speckle elimination, the value of pixel in filtered image is determined by the gain factor $(k_{(i,j)})$. It is assumed that the noise in image is unity mean multiplicative noise. If the captured noisy image is z, true image is x and noise is n, then the noisy image model can be expressed as

$$Z_{(i,j)} = x_{(i,j)} \cdot n_{(i,j)} \tag{1}$$

where the $\bar{x}$ is local mean, the speckle free pixel value is calculated by

$$\hat{x}_{(i,j)} = \bar{x} + k_{(i,j)}(z_{(i,j)} - \bar{x}). \tag{2}$$

The Lee filter tries to minimize MSE between $x_{(i,j)}$ and $\hat{x}_{(i,j)} * k_{(i,j)}$ and the gain factor $k_{(i,j)}$ is calculated by Eq. 3.

$$k_{(i,j)} = Var(x)/(\bar{x}^2 \sigma_n^2 + Var(x)) \tag{3}$$

where, $Var(x)$ is the local variance. The coefficient of variation, σ_n gives the knowledge of ratio of standard deviation to mean i.e. $\sigma_z/\bar{Z}$ over homogeneous areas of noisy image.

Another phenomenon in acoustic images is the acoustic reverberation and multipath problem which generate effects such as ghost objects. In this work, the received signal is analyzed and the homomorphic deconvolution method is applied as a means of combating the multipath problems. Followingly, the image captured by the FLS $g(x, y)$ is decomposed into the reflectance function $r(x, y)$ and the illumination intensity $i(x, y)$ using $g(x, y) = i(x, y).r(x, y)$. Using the log of the image, the components of illumination and reflectance can be separated. The log-image is Fourier transformed and High-pass filtered using the H modified Gaussian filter to Eq. 4.

$$G(w_x, w_y) = H(w_x, w_y) \cdot I(w_x, w_y) + H(w_x, w_y) \cdot R(w_x, w_y) \tag{4}$$

Inverse Fourier transform is applied to return into the spatial domain to get the filtered image.

All the filters applied up to now blurs the acoustic image. Therefore, it is important to apply some mechanism for image enhancement to strengthen some features. In this paper we tested two methods DSR and CLAHE. The CLAHE method has been described fully in [3] and therefore we refer to this article for further details. These methods have proven to be suitable for enhancing both the grayscale and colored images. The principle of the DSR is described in [5] that if optimum noise is added with the weak input signal it boosts the signal considerably and gives better signal to noise ratio (SNR).

Step 2: Sonar image segmentation, feature extraction and annotation . As objects suspended in water reflect acoustic waves more than the water environment, they are characterized by high-intensity regions on the images. Therefore, the approach for segmentation is to distinguish and separate the objects from the background [1, 4]. Due to this fact, an approach based on the acoustic image formation to detect peaks of intensity is adopted as in Santos et al. [1]. Briefly, a sonar image is composed of beams B and bins. Therefore, every acoustic beam of the acoustic image is analyzed individually for every bin. The average intensity $\bar{I}(b, B)$ is calculated for each bin b of a given beam B by Eq. 5.

$$\bar{I}(b, B) = 1/(win_{sz}) \sum_{(i=b-win_{sz})}^{b} I(i, B) \tag{5}$$

where win_{sz} is the window size, in number of bins, admitted in the averaging; b and i are bin identifiers; $I(i, B)$ is the intensity of ith-bin of Bth beam. The intensity $I_{peak}(b, B)$ is an offset of $\bar{I}(b, B)$ as shown in Eq. 6

$$I_{peak}(b, B) = \bar{I}(b, B) + h_{peak} \tag{6}$$

where h_{peak} determines the minimum height of a peak of intensity. A sequence of bins with an intensity $I(b, B)$ greater than $I_{peak}(b, B)$ are considered part of a peak and are not considered on the $\bar{I}(b, B)$ computation.

Along the beam B, the bin b_{peak} with the greater intensity $I(b_{peak}, B)$ is adopted to build the segmentation parameters. Fig. 2b shows in red values of $\bar{I}(b, B)$, in blue values of $I(b, B)$ and in green values of $I_{peak}(b, B)$ of all bins of a single beam B. The peaks detected b_{peak} are represented by colored circles. After the detection of all peaks, a neighboring search for connected pixels is performed for each peak. The 8-way neighborhood criterion is adopted by the BFS algorithm. All the connected pixels are visited if $I(i, j) > \bar{I}(b_{peak}, B)$.

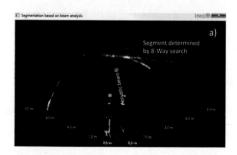 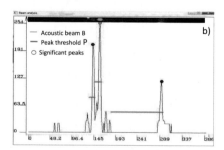

Fig. 2. a) Peak analysis of a single beam and (b) beam intensity profile

To be able to apply SVM, features need to be extracted from the segments. So after segmentation, a Gaussian probabilistic function is applied and shape descriptors for each segment are calculated (see Fig. 3a). Using the Singular

Value Decomposition (SVD), the eigenvalues and eigenvectors of the covariance matrix are computed, from which the largest eigenvalue and the second largest eigenvalue are used to define the width and the height, respectively. In addition, other shape descriptors can be calculated starting with the segments area which is computed using the Green's theorem, the convex hull area and the perimeter, the inertia ratio, the convexity, the mean and the standard deviation of the acoustic intensity of each segment. Almost all data are geometrical information, however the mean and the standard deviation of the intensities represents the acoustic data. After the Gaussians are automatically calculated, a manual segment annotation process is conducted (Fig. 3b).

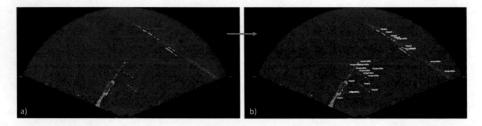

Fig. 3. a)Acoustic image gaussians and b) annotated acoustic image

Step 3: Segment classification using supervised learning . After the description and the annotation of the segments, they are now ready for classification. Well-known supervised classifiers such as Support Vector Machine, Random Trees and K-Nearest Neighbors can be used for this purpose. The OpenCV implementation of the supervised classifiers were used on this work. Four classes of objects available in our dataset (aluprofile, obstacles(1-3)) were adopted for learning.

Step 4: Generalized Fuzzy Hough Transform . After finding all the possible positions of the aluprofiles, the next step is to use GFHT for localizing the docking station in the acoustic images using its template. For detailed description of GFHT we refer to Suetake et al. [8]. In GFHT, the fuzzy concept is introduced to the voting process. Therefore we consider the area C_k containing the feature points, with radius R_c (pixel) from the point (x_c, y_c) in question. In order to consider the effect of the neighboring feature points on (x_c, y_c) in the area C_k, membership value is given to each point in the area C_k, according to the distance from (x_c, y_c). The following is the vote value in the voting process of the GFHT: Vote value $= \sum$ membership value of the feature point in C_k. This means that the effects of the feature points around (x_c, y_c) are counted up in voting.

3   Experimental results

The experimental results were obtained using real acoustic images collected from a test basin in which the docking station was installed. A video was taken using a 2D Forward looking sonar mounted on an AUV. From the video, 16-bits gray scale images with a resolution of 1429x768 were generated. The AUV dives to the level of the underwater docking station starting from different position of the test basin, left, right, center and moves towards the docking station while recording data.

Using the pipeline described in Fig. 1, the images went through filtering and enhancement, segmentation and finally manual annotation. The training data comprises a total of 627 segments over 33 acoustic frames manually classified into mainly two classes (aluminum profile = 330 and obstacle). The obstacle class is divided in to three categories according to their shapes to cover everything which is not an aluminum profile (Obst1=33, Obst2 = 165 and Obst3 = 99). The shape of the segment is the most distinctive feature for recognition. Therefore, the annotation for the classes was performed accordingly, the class obstacle 1 are the largest segments; the class obstacle 3 are the smallest segments and the aluminum profiles are small and the most convex segments. Overfitting is avoided by choosing the supervised classifier parameters (C, gamma for the SVM) using 5-folds cross validation. The folds are applied repeatedly and the average accuracy is used to choose the best parameters. Furthermore, normalization is required so that all the inputs are at a comparable range. For segmentation, the parameters such as the separation distance allowed between segments were defined empirically in several trials.

The best result was obtained using the SVM classifier with radial basis function kernel and $\nu = 1.442$ and $C = 11.739$. This Classifier reaches a hit rate of 98% and 93% for training and validation, respectively. In Fig. 4a, the ellipses in red are automatically detected by the segmentation algorithm, and the yellow labels have been manually defined. After running the classifier training, the labels in magenta, red (incorrect) or green (correct) appear to represent the classification assigned by the classifier. The magenta labels indicate segments without annotation to compare. The performance of the detection and localization system using GFHT is measured using the detection rate, i.e., the total number of detections compared to the actual docking station position in all images and the localization accuracy, i.e., the correct location detection of the docking station compared to its actual location in an image for all 674 images. The detection rate is quite high for all images with docking data achieved above 80% detection rate. Further, the system is able to localize the docking station on all sonar images quite accurately with an average position $error = 3cm$ and average 2D orientation $error = 3.9°$.

4   Conclusions and future work

A method to automatically detect and localize a docking station in acoustic images is proposed. The acoustic image is automatically segmented and the shape of each segment is described geometrically and argumented by its acoustic intensity reflected by the object. The object classification is performed by SVM classifier. The image segments are manually annotated for training. The results

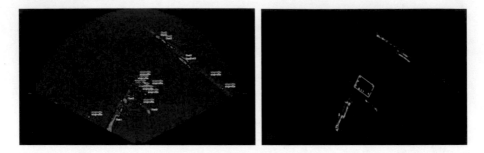

Fig. 4. a) Example results of classification and b) docking station localized by GFHT

show that it is possible to identify and classify objects such as aluprofiles in real underwater environments. With the known positions of the aluprofiles, GFHT can be performed much faster for template matching of the real docking station. From the GFHT, the 2D position and orientation of the docking station is obtained and can be used by the homing algorithm. Future works will be focused on exploring and make comparisons with end-to-end deep learning.

5   Acknowledgements

The data described in the paper were collected by students based at the Fraunhofer IOSB-AST in Ilmenau. We thank these students for their hard work during data collection, especially Nguc Pham Tham.

References

1. Santos, M., Drews, P, Núñez, P., Botelho, S.: Object Recognition and Semantic Mapping for Underwater Vehicles Using Sonar Data, Journal of Intelligent and Robotic Systems, volume **91**(2) pp 279–289. (2018)
2. Cho, H., Pyo, J., Gu, J., Jeo, H., Yu, S.C.: Experimental results of rapid underwater object search based on forward-looking imaging sonar. In: Underwater Technology (UT), 2015 IEEE, pp. 1–5 (2015).
3. Hitam, M. S., Yussof, W., Awalludin, E. A. and Bachok Z.: Mixture contrast limited adaptive histogram equalization for underwater image enhancement. IEEE, (2013).
4. Machado, M., Ballester, P., Zaffari, G., Drews-Jr P., and Botelho S.S.C.: topological descriptor of acoustic images for navigation and mapping in IEEE 12th Latin American Robotics Symposium LARS, pp. 1–6. (2015)
5. Deole M.T., Hingway S.P., Suresh S.S. Dynamic stochastic resonance for low contrast image enhancement. IOSR JVSP. **4** 1–5.(2014)
6. Evelin S. G., Lakshmi Y.V.S., Wiselin J.G.: MRI Brain Image Segmentation based on Thresholding, Int. Journal of Advanced Computer Research **3** (8). (2013)
7. Lee, J. S.: Digital image enhacement and noise filtering by use of local statistics. IEEE Trans. PAMI; PAMI **2**(2) 165–168. (1980)
8. Suetake, N., Uchino, E., Hirata, K.: Generalized Fuzzy Hough Transform for Detecting Arbitrary Shapes in a Vague and Noisy Image. Soft Computing. **10**. 1161–1168. (2006)

Deployment architecture for the local delivery of ML-Models to the industrial shop floor

Andreas Backhaus, Andreas Herzog, Simon Adler and Daniel Jachmann

Fraunhofer Institute for Factory Operation and Automation, IFF
Magdeburg, Germany
andreas.backhaus@iff.fraunhofer.de

Abstract. Information processing systems with some form of machine-learned component are making their way into the industrial application and offer high potentials for increasing productivity and machine utilization. However, the systematic engineering approach to integrate and manage these machine-learned components is still not standardized and no reference architecture exist. In this paper we will present the building block of such an architecture which is developed with the ML4P project by Fraunhofer IFF.

1   Introduction

The current industrial revolution challenges companies to optimize production to face scarcity resources and climate change. Initiatives like "Industrie 4.0" should accelerate the required technological innovations and already deepened the understanding about the value of data. Schnieder et al. [1] have written in 1999 that data is a required production resource. Nowadays, data becomes also part of the production result that can be used during operation.

Using data is currently a challenge during production. Even if the companies have identified the potentials of data [2], it is challenging to gain the added values. One reason is that IT-knowledge is in most cases not a key competence of companies in industrial production. To achieve benefits from data-driven machine learning (ML) it is required to analyse the data to gain information e.g. about optimizable parameters or aspects for predictive maintenance. There are multiple algorithms for ML that differ in their requirements, weaknesses and strength and experts were required to utilize ML. Currently there are increasingly software libraries, tools, and frameworks available which hide complex numerical processing, required in ML. Even though most systems try to provide easy access to ML, the outnumbering set requires some overview about the systems and expert knowledge, too. Additionally, there is currently a disagreement regarding data security. Some Frameworks allow the use of cloud-based services but this requires a continuous data-stream to the cloud. Some companies prefer to keep their data local to avoid knowledge transfer and a theoretical possible exposition of the production system.

© The Author(s) 2021
J. Beyerer et al. (Hrsg.), *Machine Learning for Cyber Physical Systems*, Technologien für die intelligente Automation 13,
https://doi.org/10.1007/978-3-662-62746-4_4

2   Aim of the presented work

We present a deployment architecture for the local delivery of ML to the industrial shop floor. The architecture is already used during operation in the industry but is still work in progress.

The architecture allows to perform ML locally and without the requirement to deliver data via public domain, even if it also could be used as cloud-service or together with cloud-services. The architecture consists of a server locally to the production unit. It manages the connectivity to the machines programmable logic controller (PLC), collects signals and bypasses signals to the ML-Module. Input signals are verified, and the ML Model is monitored; thus, it can be guaranteed that invalid results are detected. Additional server connectors provide access to available knowledge systems (e.g. ERP, PDM, PLM) and the web-based client. It provides an assistance system for workers, so they can query and collect data during daily routine. ML analytics requires that the digital information model is up to date. The web client gives workers the possibility to keep the information model up to date during their work process.

Using ML during production requires not only the collection of data for training and analysis. Production has additional requirements that are practically motivated or required by law. We therefor present and discuss the requirements raised up during conception and implementation of the presented architecture.

3   Related Work

As the ML paradigm has moved from a scientific exercise to a tool for industrial data processing, the need has arisen to export or describe ML-models independent of their initial creation framework. All large frameworks, some of them not in existents only a couple of years ago, now offer a model description format as well as runtimes for scoring the described model. A model is basically described as graph of computing steps to arrive at an output value from an input value while the training process determined the free parameters of the model in order to minimize a set loss function. As field progresses, the framework provider for training the ML model not necessarily needs to be the provider of the runtime system for scoring the model in production nowadays also referred as 'model serving'. Examples for model descriptions are xml-based file formats like PMML [3] or binary formats like ONNX [4] or Tensorflow Graphs [5].

In industrial production the company must document their machine configurations, so they can prove the operation conditions at any time. Changes during maintenance are documented with version and revision. If ML is influencing industrial production, they must be documented in a comprehensible representation, too. The advantage of model descriptions is that components of a numerical system can be described in a form that is independent of the model version of a particular learning framework. It furthermore increases the readability and transparency of a numerical model since all computing steps are described (comprehensible).

However, due to fast moving nature of the field of machine learning, not one standardized way of describing models has established yet, even when efforts are being made

[6]. Furthermore, the modules of numerical processing systems are increasingly built by a number of different participants with their favourite frameworks. Due to this fact, we argue for a generalized model serving description that can integrate numerical transformation descriptions from different sources.

4   Architecture

ML is currently discussed as integrated part of controllers from technical systems, to dynamically adapt technical systems to unknown production conditions. ML-models that decide about operation parameters due to analysed data, yield to technical problems and law-relevant questions about responsibilities. These problems can be avoided, if the operator remains responsible for the technical system but is utilizing ML as part of an assistance system. The assistance system has a server and a client part. The server is a communication hub, which connects to different knowledge bases. Each connection translates between an internal protocol and the proprietary protocol of the connected knowledge base.

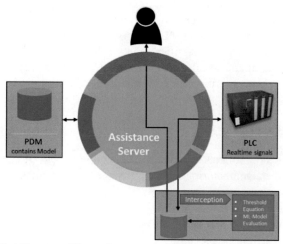

Figure 1: Simplified Structure. The assistance server is a communication hub, unifying protocols. Signals provided by plc are stored a connected database and can be analysed online. ML-models are provided as part of the machine documentation by a connected PDM system.

The machines plc provides the signal data (sensor, actor, states) and is one knowledge base. Like MES, the assistance server monitors the signal data and stores the data into internal databases, but can intercept signal-streams for online analyses. The resultant values are stored as additional soft sensor. While simplest implementations of such an interception are thresholds or equations, ML methods allow evaluating signal values against trained models.

The machines signals will change their behaviour over time e.g. of wear and tear. Therefore, the models must adapt and may only be conditionally valid. It is very likely that during failure, signals arise that never were trained and are not part of the model.

Even if ML-models can extrapolate, they lose accuracy. Especially during critical operational states (unplanned failure) the machines operator must understand the trustworthiness of the models results.

Therefore, the ML-models requires monitoring on different levels. On the first level, the range and statistical parameters of input and output signals must be checked individually. If an input signal is out of range, because of sensor or connection fault, the system must warn and switch to a save mode. The save mode can be an alternative ML-model, which does not considers the input value; a more robust model, which persevere a single sensor failure, or if no more options exist, a controlled machine stop. On output side, the ML-model does not know the limits of the machine actuators (linear axis, motors, oven, etc.). To prevent the actuators from being destroyed, output signals must be verified too.

On second level, we consider that the training of ML-models consists only a part of the possible combinations of input values. All input combinations outside the training vectors may result in undefined model output. A one-class classifier [Pimentel2014] can be used to detect such an input.

The third level controls the stability in the current working point of the ML-model. To estimate the stability a small variation to the current input value (working point) is added and the model output is processed. Depend on the calculation speed, this can be done in time multiplex or on parallel ML-models. If the output variation exceeds a predefined range, the model can be classified as unstable.

All monitoring components added to the plain ML-model result in an industrial runtime with parameters described in an overall model description.

5   Data connectivity and collection

The assistance server collects signal data from connected plc. It can provide parts of the functionality of a manufacturing execution system (MES) but allows also analysing the signals data online.

Connection and signal collection may be difficult. PLC have the major task to execute the programmed logic. Sending (push) actively data from the PLC is a low priority task and is not executed if the plc logic requires resources. Alternatively, the assistance server can scan (pull) the PLCs memory for changes. No changes are required in the PLC, but the update frequency must be higher than the Nyquist frequency of the fastest signal monitored. Besides the signals used to operate all machine part, both approaches have the benefit that internal flags and error codes can be monitored additionally.

For larger machines, a bus-logger can mirror signals from the communication bus and sends signal values if they pass the bus. Because signals are not taken from the PLC, one must be aware that only control required signals are detected. Error signals are only send to the connected HMI, but not distributed over the communication bus. For ML analysis, it is important to know which signals are related to malfunctions or regular operation states. Using bus mirrors requires therefore querying data from multiple sources and synchronization strategies.

For small and medium enterprises, pulling signals can be the easiest and cheapest way to collect operational data. The number of monitored signals can be limited, if expert knowledge is considered. Experts can select relevant sensors. The number of signals can be reduced if dependencies between the signals are detected using ML. Reducing the number of signals allows a higher scan rate to pull sensor data. Additionally, it should be questioned which sensor values are relevant to describe the process. Data analysis provides information to influence some kind of reduced if dependencies between the signals are detected using ML. Reducing the number of signals allows a higher scan rate to pull sensor data. Additionally, it should be questioned which sensor values are relevant to describe the process. Data analysis provides information to influence some kind of process. Controlling a robotic system is a high frequency process while a logistic process is even slower and allows a lower update rate.

This is not true in general. Even in slow processes, high frequency events can be an indicator for important effects, but experts are most likely aware of this and can adjust signal capture strategies.

6   ML-Model Serving

For a reliable integration of machine learning models into the industrial application, a model description must be generated from the training framework and contain a description of all necessary numerical processing steps. Therefore, a general model description and a model-serving component is needed. In Figure 2 the classical ML-model description (a) is extended to incorporate the ideas and building blocks described below.

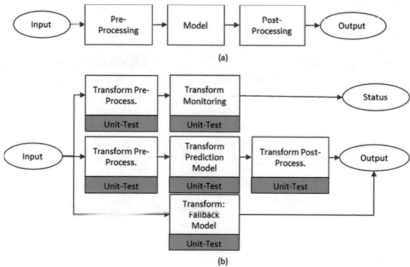

Figure 2: In its current form, a machine learned model is described as a single pipeline of pre-processing, model inference and subsequent post-processing (a); for industrial deployment, this pipeline must be generalized into a graph of transforms which describe the model deployment as a combination of the prediction model combined with monitoring and fallback models as

well as a number of transform for example for pre- and post-processing (b); each transform is stored with a unit test in order to check at runtime whether the runtime implementation is correctly calculating the transformation steps.

We can formulate the following requirements for this description:

a. Constancy: The implementation of a model serving should not change when a new model is deployed. Patching and updating of at machine serving component needs to be avoided. Saving parameters only is not sufficient, because a full description of the transformation function in the model description is required.

b. Flexibility: Along the machine-learning pipeline developed in ML4P, different participant should be given a number of supported modelling frameworks to describe the base models (or base transforms) in the serving structure, since one will not find one description or framework that will provide all functionality. Furthermore, core transformation descriptions should be exchangeable since the market of machine learning frameworks is in constant movement.

c. Data Interpretability: The serving description must incorporate Meta information about the expected input and output data. Each base model needs a Meta description too, since different partners along the ML4P pipeline generate them.

d. Transform Testability: No transformation is allowed into serving without passing a unit test with attached I/O test data, per transform.

7   Monitoring Strategies

The basic principal of machine learning is to model a system behaviour by observing the system and fitting a generic mathematical model to the observed data by means of numerical optimization. In comparison to a physical model, a machine learning model is only valid within the boundaries of the dataset it was trained on. A physical model is valid everywhere the laws of physics apply. This arises to the need to monitor the input and output data of an ML-model as well as to observe its behaviour. Figure 3 shows monitoring strategies. Monitoring is performed by one-class models that act as anomaly / novelty detectors [8].

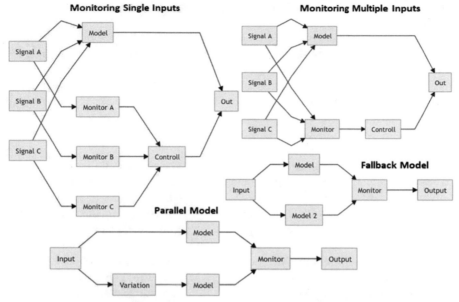

Figure 3: Several monitoring strategies, which can be deployed to check whether a model is still working in a valid data environment.

1. **Monitoring single inputs:** A model input can originate from different sources, which should be modelled for their valid behaviour separately. This strategy is for spotting abnormalities in sensor inputs. A similar control system can monitor the model output to prevent actuator damages.
2. **Monitoring multiple inputs:** A monitoring model is checking the validity of the complete input vector in order to determine whether the input data is still within the boundary of the training data.
3. **Fall-back Model:** In case of an abnormality, a fall-back model is added to the model description in order to continue operations. A fall-back model could be a physical model, or a more transparent less parameter machine learning model.
4. **Parallel Model:** The input is processed with a parallel copy of the prediction model. The input is varied with in magnitude of the signal noise and used to measure the prediction model stability at the model's working point.

8   Lifecycle Management

ML Models are stored related to the structure of the technical system. A technical system has different types of bill of materials (BOM) that depict structure how parts of the technical system are connected in a specific view. In manufacturing the engineering and service BOM are motivated by spatial view, because they have their origin during engineering. A BOM is a very common type of an ontology how parts of a technical system are related.

ML is used to analyse the behaviour of a technical system, but behaviour dependencies are currently not formalized very often. Therefor additional ontologies like SysML [9] may become a higher importance in the future.

Maintenance is the process to keep a technical system alive and may require replacing parts during the product life cycle. The replacement is most likely not identical to the original part. It maybe new and without wear and tear, a newer version, from a different distributor, or a completely new setup. What does that mean for a trained ML model?

After replacement the operation of a ML model must be considered as on trial. Using the monitoring, the ML model can provide a confidence value how much the operator may trust the values. This also allows reusing older models if models adapt during operation. After replacement with an identical part, the current ML model and an older version, which was used in the beginning of the parts life cycle, can be benchmarked against each other.

For a larger machine, multiple ML models may be used to for different functionalities of the machine. The operator requires some overview where he can trust the data driven decision support using ML. This requires that it is documented which signals of which parts are used as input and which parts, assemblies, or modules are influenced by the ML model result. As an example: Consider that there are cameras that monitor the heating of a material. The cameras maybe part of an assembly but the material is in a heater assigned to a different assembly. The ML model is using the image data to monitor the heating process. If a camera is replaced it may influence the utilized image data. From a technical point of view, the model is using data of the camera's assembly, but the effects will influence the decision making of the material heating.

The ML-model is associated with the parts from which it uses signal values. If a part is changed during maintenance, every model can be identified that may be compromised. Additionally, the association to all influenced parts is required, to inform the operator automatically for which part and function the derived result is temporarily invalid.

9   Discussion

Using ML to optimize production requires interdisciplinary work between ML experts and experts from industry. A sustainable integration of ML models requires the integration of models in the product life cycle and their maintenance. Operational personal must be able to understand not only results of the online analyses but also their dependencies and their trustworthiness.

We established the corner stones of a deployment architecture for data-driven models acquired by machine learning in an industrial machine environment. While products like personal assistants that incorporate some form of machine learned component are now distributed widely, not much standardization on the engineering process of deploying and managing machine learning models has been formulized. Here, the Fraunhofer project "Machine Learning for Production (ML4P)" will serves as a testing ground for implementing the deployment methodology laid out in this paper.

10   Acknowledgement

This work was supported as a Fraunhofer LIGHTHOUSE PROJECT.

References

[1] Eckehard Schnieder: Methoden der Automatisierung. Vieweg+Teubner Verlag, 1999.
[2] M. Eisenträger, S. Adler, M. Kennel and S. Möser, "Changeability in Engineering," 2018 IEEE International Conference on Engineering, Technology and Innovation (ICE/ITMC), Stuttgart, 2018, pp. 1-8.
[3] Specification PMML Format Version 4.3: http://dmg.org/pmml/v4-3/GeneralStructure.html
[4] Specification ONNX Format: https://github.com/onnx/onnx
[6] Martín Abadi et. al TensorFlow: Large-scale machine learning on heterogeneous systems, 2015. White Paper https://www.tensorflow.org/about/bib
[7] Sculley, D & Holt, Gary & Golovin, Daniel & Davydov, Eugene & Phillips, Todd & Ebner, Dietmar & Chaudhary, Vinay & Young, Michael & Dennison, Dan. (2015). Hidden Technical Debt in Machine Learning Systems. NIPS. 2494-2502.
[8] A review of novelty detection. Marco A.F.Pimentel, David A.Clifton,Lei Clifton,Lionel Tarassenko. Signal Processing, Volume 99, 2014, Pages 215-249
[9] Matthew Hause, The SysML Modelling Language, Fifteenth European Systems Engineering Conference, 2006

Deep Learning in Resource and Data Constrained Edge Computing Systems

Pranav Sharma, Marcus Rüb, Daniel Gaida, Heiko Lutz, and Axel Sikora

Hahn-Schickard-Gesellschaft für angewandte Forschung e.V.,
Wilhelm-Schickard-Str. 10, 78052 Villingen-Schwenningen, Germany
{firstname.lastname}@hahn-schickard.de

Abstract. To demonstrate how deep learning can be applied to industrial applications with limited training data, deep learning methodologies are used in three different applications. In this paper, we perform unsupervised deep learning utilizing variational autoencoders and demonstrate that federated learning is a communication efficient concept for machine learning that protects data privacy. As an example, variational autoencoders are utilized to cluster and visualize data from a microelectromechanical systems foundry. Federated learning is used in a predictive maintenance scenario using the C-MAPSS dataset.

Keywords: Variational Autoencoders · Federated Learning · Unsupervised Learning · Predictive Maintenance

1 Introduction

Usually, deep learning methods are in need of a lot of labeled training data and computing resources to exploit their full potential. In most industrial applications, labeled training data is very expensive and time-consuming to collect. With the ongoing trend of bringing artificial intelligence (AI) on edge and embedded devices, also known as edge AI, the computational power is limited too. In this paper, methodologies that counteract the scarcity of labeled data are presented and exemplified by selected applications from production industry. These methods are variational autoencoders and federated learning [3], which are applied to the following applications:

1. Clustering and visualization of wafermap patterns
2. Anomaly detection for sensor data of a furnace
3. Predictive maintenance using federated learning

In the first two applications unsupervised learning is employed, which is a classical methodology to detect patterns in data without the need of labeling the data. In the latter application, federated learning is used to demonstrate its use in the case of edge AI. All of these applications are used as examples, to demonstrate the usage of aforementioned techniques.

J. Beyerer et al. (Hrsg.), *Machine Learning for Cyber Physical Systems*, Technologien für die intelligente Automation 13,
https://doi.org/10.1007/978-3-662-62746-4_5

2   Methods & Related Work

In this section, variational autoencoders and federated learning are introduced.

2.1   Variational Autoencoder

Autoencoders belong to the family of unsupervised machine learning methods and are used for dimensionality reduction. An autoencoder encodes high-dimensional input data to a lower dimensional latent space and then decodes this back to its original dimension to restore the input data. A variational autoencoder (VAE) encodes the input to corresponding mean and variance, which means that the input data is assumed to come from (or) generated from a statistical process [2]. These mean and variance are used to reconstruct the input during training. Doing this, forces the encoding of the latent space to be meaningful everywhere. For both methods, the lower dimensional latent space is used to e.g. analyse or visualize the original data distribution.

2.2   Federated Learning

The AI market is dominated by tech giants like Google, Amazon and Microsoft, which provide cloud solutions and APIs (application programming interface) for AI. This monopolization of data, develops mistrust, especially in small and medium-sized companies to make their data available for AI or to use it themselves. Instead of collecting data and sharing it in a data center, the data should be kept on the embedded devices where it is collected. To be able to use AI in this scenario, McMahan et al. [3] introduced a learning algorithm in 2017 that allows any number of clients with local training to improve the model parameters of a global model shared with all other devices. This algorithm is called federated learning that follows the approach of *"bringing code to data instead of data to code"*.

Imagine a production chain in which several motors and heating elements are in operation. In order to avoid production downtimes, the machines are equipped with sensors that allow to do condition monitoring. Predictive maintenance algorithms estimate the next maintenance date based on the results of this monitoring. It is evident that such sensitive production data should not leave *"the house"*. To prevent this from happening, the machine learning model is trained with the locally kept data in the company and only the model parameter changes are forwarded to the server. The server collects the parameters of each production line and aggregates them by the federated averaging algorithm. The updated model is then redistributed to all clients.

The use of federated learning in the application case of mechanical manufacturers has some differences compared to the original intended application by Google. The main difference is the number of clients and therefore less possibility to compensate for outliers.

3   Results

3.1   Clustering and Visualization of Wafermap Patterns

(a) Input   (b) Reconstruction

Fig. 1: Comparison of input wafermap image with reconstructed wafermap image.

Production of chips from silicon wafer requires optimum performance checks for each chip, which are typically electrical measurements. The electrical measurements for all chips of a wafer results in a wafermap (see Fig. 1). A wafermap visualizes the measured values of one electrical measurement as color-coded values. Wafermaps produced in a production process, may contain patterns that are result of production and material changes over a time horizon. The observed patterns can be utilized to gain insight into their cause of production.

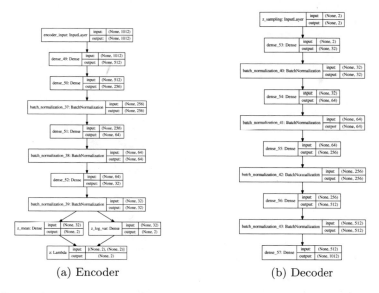

(a) Encoder   (b) Decoder

Fig. 2: Structure of encoder and decoder section of wafermap VAE.

The wafermaps studied in this paper come from the microelectromechanical systems (MEMS) foundry at Hahn-Schickard. A diverse range of patterns are available in these wafermaps. This requires identification of patterns on all of the available wafermaps. To avoid to have to go through all of the wafermaps manually and keep track of all the patterns along with adding labels, we chose unsupervised learning techniques to help cluster all wafermaps with similar patterns in the same cluster.

Fig. 3: Visualization of representations for the learned latent space.

Following [4] and [6], it was chosen to use a variational autoencoder to learn a lower dimensional latent space for the wafermaps of the available wafers. A two dimensional latent space representation was utilized, to make the encodings more human interpretable. This lack in dimensionality led to a bias in reconstruction of the wafermaps. This bias can be seen in Fig. 1b. But this wafer reconstruction did not change the patterns appearing in the wafermaps. The architecture of encoder and decoder subsection of the variational autoencoder is shown in Fig. 2a and Fig. 2b, respectively. One can view the latent space based reconstructions produced by trained variational autoencoder for wafermaps in Fig. 3.

Fig. 4: Clustering of encodings formed in the latent space using k-means clustering and visualization of wafermaps that are representative for their clusters (best viewed in color).

Once the encodings are generated, they are utilized to perform clustering of patterns available in wafermaps. A k-means clustering method was utilized to identify the clusters in the given latent space as seen in Fig. 4. From Fig. 4 one can see how many different patterns there are and how often each pattern appears.

With this information, one can deal with the most frequent patterns and try to identify the processes that produce these patterns, to avoid the patterns in the future.

3.2 Anomaly Detection for Sensor Data of a Furnace

Fig. 5: Raw visualization of latent space encodings.

Fig. 6: Visualization of latent space encodings with fitted Bayesian distribution.

Measurements of various sensors (eight temperature sensors, a couple of gaseous concentrations, timestamps, etc.) were recorded during the manufacturing process in a furnace. Proper detection of anomalies in this recorded high dimensional space of timeseries data is difficult and error prone. To deal with this, unsupervised deep learning was used for dimensionality reduction.

The data of such a process has to be processed properly, as it is time dependent. First, a difference between all of the consecutive datapoints of all measurements (including time) is calculated and is appended to the state as input too. Then all data is normalized.

In this paper, a variational autoencoder was utilized to reduce the multi-dimensional process parameters to a two dimensional latent space as shown in Fig. 5. The architecture of encoder and decoder section of the VAE can be seen in Fig. 7a and Fig. 7b, respectively. Once the network is trained, one can see cluster of points for all process steps, such as heating up, processing and cooling

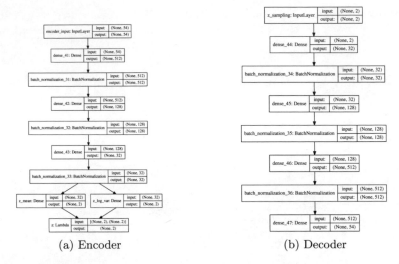

(a) Encoder   (b) Decoder

Fig. 7: Structure of encoder and decoder section of VAE for sensor data of furnace.

down. If one sees the need, then the encoder network can be further tuned to separate the clusters even more by means of labeled process information.

Once the final latent encodings for all different process steps are produced, one can fit them into individual bayesian mixture distributions, see Fig. 6. Then, distance from the observed distribution is used to detect anomalies. Fig. 8 shows a well performing batch, which can be observed via the distance from observed distribution.

Fig. 8: Visualization of a well performing batch for the process.

3.3 Predictive Maintenance using Federated Learning on Edge Devices

With the use of federated learning, a use case is presented with a machine manufacturer offering a predictive maintenance service to its customers. Each customer updates a machine learning model using the data of its machine and sends the model update back to the machine manufacturer. The developed network ar-

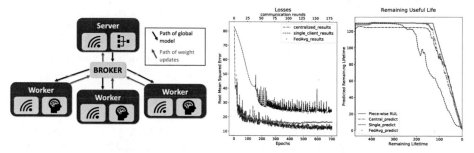

(a) Federated learning architecture   (b) Comparison of learning results

Fig. 9: Architecture for federated learning; Comparison of learning results for federated learning, one big centralized dataset and one worker standalone, respectively.

chitecture is shown in Fig. 9a. The architecture can consist of a large number of workers and one server. Workers are edge devices at different performance levels, which are installed at the different customer locations, where they remotely train neural networks on the respective data. Here Nvidia Jetson Nano, Raspberry Pi 4 and Intel NUC were used as workers and Raspberry Pi 3 as server. All workers consist of a neural network model and a transmitter/receiver unit that controls the transmission and reception of shared data. The server consists of an aggregator part that combines the individual weights in a global model and a sender/receiver which connects the server to the network.

MQTT is commonly used as network protocol in the field of internet of things (IoT) due to several reasons. Among others it has small overhead, it can be scaled easily, it is easy to implement and preserves privacy, thus it is used in this federated learning scenario as well.

Till date, we have tested federated learning in a network with four to 30 workers on predictive maintenance datasets. As a result, each subdataset of the well-known C-MAPSS dataset [5] was distributed unevenly among these workers. To improve the federated learning results on each subdataset, the learning rate of the optimizer was adjusted as well as the number of learning epochs on the customer data. Taking the training on a centralized storage as benchmark with the complete training dataset stored at one place, the federated averaging (FedAvg) [1] algorithm performs very well. Fig. 9b shows losses from training of centralized model and the single worker model over the learning epochs. The loss incurred for federated learning is shown above the *"communication rounds"* axis. A single *"communication round"* contains the four steps: (1) Distribution of the model from server to workers, (2) Training of models on worker's data for several epochs, (3) Sending weight updates from workers to the server, and (4) Aggregating weight updates by federated averaging into a new global model.

Fig. 9a shows that learning on distributed datasets achieves almost the same accuracy as learning centralized. This comparison is made only to assess the performance in terms of prediction. Centralized learning can only be seen as a benchmark, not as a real alternative as privacy preservation is not sufficient. The learning results of federated learning, when compared with results attained by a single worker (without connection to federated learning), are better, because one single worker operates on less amount of data. The amount of data available to one single worker, was often not sufficient to achieve clean convergence.

4   Conclusion

In this paper, it was shown that deep learning can be used for industrial processes, where data is scarce and data privacy is important. Mainly, unsupervised learning methods such as variational autoencoders can be used to cluster and visualize high-dimensional data as was shown with clustering of wafermaps and visualization of process data of a furnace. For a sequential and fixed duration process, one can use such clustering of encodings for monitoring the undergoing process. Also, for monitoring the final production of any process, latent space encoding of samples can give insight into issues and opportunities for the process.

Furthermore, the basic ideas of federated learning were introduced, which make them predestined for industrial use cases. It was shown that the accuracy of the predictions from a federated learning model is in a similar range to the prediction results of centralised training, based on same inputs. In federated learning, any increase (with required data privacy) in the data volume should thus lead to an increase in the quality of predictions. The customer of a machine with a predictive maintenance model, that is optimized via federated learning, will benefit by reducing production downtime through intelligent algorithms.

References

1. Bonawitz, K., Eichner, H., Grieskamp, W., Huba, D., Ingerman, A., Ivanov, V., Kiddon, C., Konecny, J., Mazzocchi, S., McMahan, H.B., Van Overveldt, T., Petrou, D., Ramage, D., Roselander, J.: Towards Federated Learning at Scale: System Design (2019), http://arxiv.org/abs/1902.01046
2. Kingma, D.P., Welling, M.: Auto-Encoding Variational Bayes. In: *ICLR 2*. p. 10 (2014)
3. McMahan, H.B., Moore, E., Ramage, D., Hampson, S., Arcas, B.A.y.: Communication-Efficient Learning of Deep Networks from Decentralized Data. In: *AISTATS*. vol. 54, pp. 1273–1282 (2017)
4. Santos, T., Kern, R.: Understanding wafer patterns in semiconductor production with variational auto-encoders. In: *ESANN*. pp. 141–146. No. April, Bruges, Belgium (2018)
5. Saxena, A., Goebel, K.: Turbofan Engine Degradation Simulation Data Set (2008), http://ti.arc.nasa.gov/project/prognostic-data-repository
6. Tulala, P., Mahyar, H., Ghalebi, E., Grosu, R.: Unsupervised Wafermap Patterns Clustering via Variational Autoencoders. In: 2018 International Joint Conference on Neural Networks (IJCNN). vol. 2018-July,

pp. 1–8. IEEE (jul 2018). https://doi.org/10.1109/IJCNN.2018.8489422, https://ieeexplore.ieee.org/document/8489422/

Prediction of Batch Processes Runtime
Applying Dynamic Time Warping and Survival Analysis

Paolo Graniero[1] and Marco Gärtler[2]

[1] Università La Sapienza, Rome, Italy
`paolograniero8@gmail.com`
[2] ABB AG, Corporate Research Germany, Ladenburg, Germany
`marco.gaertler@de.abb.com`

Abstract. Batch runs corresponding to the same recipe usually have different duration. The data collected by the sensors that equip batch production lines reflects this fact: time series with different lengths and unsynchronized events. Dynamic Time Warping (DTW) is an algorithm successfully used, in batch monitoring too, to synchronize and map to a standard time axis two series, an action called alignment. The online alignment of running batches, although interesting, gives no information on the remaining time frame of the batch, such as its total runtime, or time-to-end. We notice that this problem is similar to the one addressed by Survival Analysis (SA), a statistical technique of standard use in clinical studies to model time-to-event data. Machine Learning (ML) algorithms adapted to survival data exist, with increased predictive performance with respect to classical formulations. We apply a SA-ML-based system to the problem of predicting the time-to-end of a running batch, and show a new application of DTW. The information returned by open-ended DTW can be used to select relevant data samples for the SA-ML system, without negatively affecting the predictive performance and decreasing the computational cost with respect to the same SA-ML system that uses all the data available. We tested the system on a real-world dataset coming from a chemical plant.

Keywords: Batch Process Monitoring · Dynamic Time Warping · Time-to-end Prediction · Survival Analysis

1 Introduction

Batch production is a common manufacturing technique for chemical, pharmaceutical and food industries where a given product is realized in a stage-wise manner following a given formula or recipe. Such recipe defines rigorously the sequence of steps that each batch run follows. Nonetheless, it is quite common that distinct batch runs for the same recipe differ in several aspects, some of them time-related. For example, total duration or duration of the individual subphases can vary and, even in synchronized phases, significant events or alarms can happen at different relative positions in time. The magnitude of such variations requires advanced process monitoring techniques to ensure the consistency

© The Author(s) 2021
J. Beyerer et al. (Hrsg.), *Machine Learning for Cyber Physical Systems*, Technologien für die intelligente Automation 13,
https://doi.org/10.1007/978-3-662-62746-4_6

and the quality of the product, to improve the safety levels of the plant, and to better understand and control the process [16]. This monitoring is, in turn, beneficial to the process operations, planning and can lead to overall improvements. A particular branch of process monitoring is statistical process monitoring: these techniques rely on mathematical tools that help to identify and control variations in the production process, analyzing data coming from the reactors. This data, collected from a large number of sensors, comes with some peculiar characteristics as a consequence of the temporal variability stated above, affecting the analysis performed. Many established statistical techniques for process monitoring, such as Multiway Principal Component Analysis (MPCA), require that the input data have equal length, that for a batch process is its duration. As stated above, even in case of the same duration, significant events could be asynchronous, leading to problems during the analysis, such as comparing different but synchronous events. These issues are not unique to batch data: Dynamic Time Warping (DTW), originated in signal processing [19], is a widely successful and often adopted technique to homogenize data with a different time frame to a standard duration and synchronization of characteristics. An adoption to batch data is given, for example, in [9].

There exist different versions of DTW. In batch monitoring, standard DTW is used to align two completed batches, mainly during offline analysis when completed batches are available. The open-ended version is useful in online scenarios, where a running, and therefore incomplete, batch is aligned to a prefix of a completed batch. As such, a comparison between the running batch and a historical one is straightforward. Unfortunately, such alignment does not give any information on the remaining time frame; for example, the time left before completion of the batch, the focus of our work. This specific issue can be seen as a time-to-event modeling problem, addressed by techniques such as Survival Analysis (SA). SA is a standard adopted technique in clinical studies to model the time until the occurrence of an event of interest, usually linked to the course of an illness. Many machine learning (ML) algorithms, like the ones described in [6], have been adapted to handle survival data, resulting in improved predictive performance over more standard techniques.

In this work, we propose a system combining DTW and SA that aims at predicting the total duration of a running batch given historical information on the process. We first check the feasibility of using SA in the context of batch monitoring. Then we investigate if by applying DTW, it is possible to obtain any improvement regarding the predictive performance or the computational cost. As described in Section 5, we apply DTW, not as a time normalization technique or to compute a distance measure, but as a data-selection tool: we use the mapping information contained in the warping path, one of the algorithm's output, to select only a fraction of the data available. This data is used to train a SA-ML-based algorithm that, followed by a standard regression model, returns an estimate of the time-to-end of a running batch. We show that the same SA model trained only on the fraction of data selected via DTW performs as good as the same model trained on all the data available, sometimes even better. A significant improvement is observable in the computational cost of the approach.

The rest of the paper is organized as follows. Section 2 and 3 expose some useful facts about Dynamic Time Warping and Survival analysis, respectively. Section 4 briefly describes the batch data used in evaluating the system proposed, and some data preprocessing realized. Section 5 describes in detail the proposed system, while Section 6 shows an evaluation of the results obtained. Finally, Section 7 concludes the paper.

2   Dynamic Time Warping

Dynamic Time Warping (DTW) is the name of a class of algorithms used to compare two series of values with possibly different length; the main applications regard time series. The idea behind this technique is to stretch and compress the two series to make one resemble the other as much as possible. This warping accounts for non-linear fluctuations of the time axis.

Once warped, the two series have the same length, and similar patterns are aligned. These are the time-normalization and event synchronization effects, respectively.

The main objects of interest are two: the warping path and the DTW distance. The warping path is a mapping between the time indexes of the two series. We can interpret multiple correspondences between one index on a series and multiple ones on the other as the stretches and compressions mentioned above. The warping path is the optimal mapping that minimizes the distance between the warped series: this minimal distance is the so-called DTW distance.

Usually, when using DTW, we identify a reference series and a query one. The usual set-up is to select only one reference Y to which we align several queries X^i, $i \in \{1, \ldots, I\}$.

The standard version of DTW aligns the two series in their entirety, mapping the end-points of the two series to each other. Formally, if

$$Y = (y_1, \ldots, y_N) \qquad X = (x_1, \ldots, x_M)$$

then the standard alignment satisfies the following condition for the end-points

$$x_1 \to y_1 \qquad x_M \to y_N$$

On the other hand, the open-ended version aligns a query to a reference's prefix:

$$x_1 \to y_1 \qquad x_M \to y_n, \ 1 \le n \le N \tag{1}$$

In this work, we mainly use the open-ended version, that is suited for online applications, being able to align an incomplete batch to a reference one.

DTW has been applied for time series classification, clustering, in various domains. References to such applications can be found in [5]. In these applications, the quantity of interest is mainly the DTW distance.

First applications of DTW to batch process monitoring can be found in [17,9]. The emphasis of these works is on the alignment of batch data, but only offline (completed batches). Online applications of DTW are present in the literature, for example in [4], mainly in conjunction with more advanced monitoring techniques as MPCA. In this work we use open-ended DTW as a data-selection tool: we use the index n in Equation 1 to select the points $x_{N_i}^i$ of the historical batches that were mapped to the same point. We suppose that these points are the ones containing most of the relevant information to the time-to-end prediction at the given time on the running batch.

3   Survival Analysis

Survival Analysis (SA) is a sub-field of statistics where the goal is to analyze and model the data where the outcome is the time until the occurrence of an event of interest [20]. The main feature of SA is the ability to deal with censored data, that is when the event could be unobserved in the time-frame considered in an experiment. Standard approaches to SA allow to model the time to the event of interest and obtain an estimate of it. The statistical modeling is the focus of classical approaches, while their predictive performance is limited. More recent developments focus on machine learning approaches to SA [6], adapting loss functions of known algorithms to the specific problem of SA, time until the event of interest. For survival models that do not rely on Cox's proportional hazards assumption [3], the predictions are risk scores of arbitrary scale and not the actual time-to-event. If samples are ordered according to their predicted risk score (in ascending order), one obtains the sequence of events, as predicted by the model [2]. In this work, we use one of these algorithms, Gradient Boosting Survival Analysis [6,1]; we apply a standard regression algorithm to convert the predicted risk to an actual time-to-event estimate.

4   Data

The data used to test the proposed system comes from a chemical batch production line. The data consists of 383 batches, spanning three years (2015-2017) of measurements, with a duration between 146 and 960 minutes. We represent each batch as a multivariate time series: each dimension is a process variable (PV), with a standard sampling rate of 1 value per minute for every batch and every PV. The PVs come from different sensors that equip the production line: they can represent engineering variables, control variables, or state variable. The exact nature of the PVs is confidential.

Data normalization. The application of DTW requires the data to be normalized to avoid artifacts due to different scales of the values. When applying DTW, we have to select a batch as reference, to which we align every other batch. To have a coherent online normalization, we decided to use a Min-Max scaling approach, and as the normalizing interval, we chose the range of values of each PV in the reference batch. This choice makes the normalization of the data coming from the running batch straightforward.

Reference batch and significant features. For every set of batches considered during the experiments, we had to select one batch as reference, considered the typical one. Following domain expert's advice, we chose the batch with median duration. Since constant variables have no warping information, we disregarded PVs with a constant trend in the reference, removing them from the rest of the batches too.

5   Proposed System

The system proposed in this work, which we call SA+DTW-system, is schematically represented in Figure 1. It has two phases: offline and online.

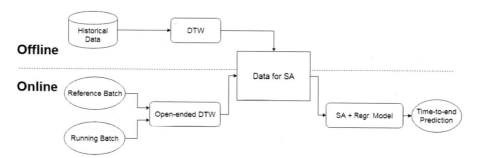

Fig. 1: Overview of the proposed system

During the offline phase, the historical data about previous batches is aligned to the selected reference batch using open-ended DTW on every prefix of the batches, storing the information about DTW distance and alignment. The resulting dataset contains the following information:

- The current time index on the running batch
- The index n of Equation 1. It's the end point of the reference's prefix to which the running batch has been aligned
- The DTW distance between the query batch and the reference's prefix
- The value of every PV of the query batch at the given time
- The time-to-end of the batch. This quantity is the target variable of the model

During the online phase, the running batch is aligned to the same reference as in the offline phase via open-ended DTW. From this alignment, we get two pieces of information: the index n of the mapped prefix on the reference and the DTW distance. n is used to filter the dataset obtained offline: only the entries mapped to the same n are selected and kept for the next step. This step consists of training a SA-ML-based model to learn the risk score based on the features mentioned above. The model is then able to assign a risk score to the running batch. This risk score is then converted to an actual time-to-end estimate by a regression model (Random Forest) trained on the risk scores assigned to the filtered dataset.

Table 1: SA-system stats

Year	2015	2016	2017
Training dataset size [rows × columns]	39431×28	77989×34	79833×34
Training time [minutes]	34	126	134
Single prediction time [seconds]	∼ 0.6	∼ 1.1	∼ 1.1

Table 2: SA+DTW-system stats

Year	2015	2016	2017
Average dataset size [rows × columns]	149×28	180×34	131×34
Single prediction time [seconds]	< 1	< 1	< 1

Software used We used the Python [18] programming language to develop the whole system. The python packages we used are: Numpy [11], Scikit-learn [12], SciPy [8], Matplotlib [7], Pandas [10], Scikit-survival [14,15,13]

6   Results

The data set had distinct characteristics in each year. Thus we split the overall data set into three parts and considered them on their own. For each year we considered approximately the first 2/3 of the batches as historical data, and the remaining 1/3 of the batches was used to test the system. In particular, we have for the three years considered the following historical/test batches: 80/40, 101/50, 75/37.

The focus of this work is on the improvements that can be obtained selecting the data via DTW, in particular before applying SA. We compared three methods to this end. The first method uses the average duration of the batches longer than the running batch as an estimate for the total duration of the running batch. This method uses only historical information, and it is considered as a minimum performance to be reached by any system to be considered useful. The second method, the SA-system, uses all the historical data at once as training data for a SA-ML based algorithm, followed by a regression model (Random Forest) for the conversion of the risk score to time-to-end estimate. All the predictions obtained by this method come from the same model. The third method is the system proposed, the SA+DTW-system described in Section 5. Figure 2 shows the results for the 2017 data. The results of the other two years show no significant differences from this one.

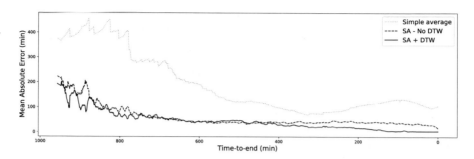

Fig. 2: Comparison of the performance (mean absolute error) of the three models tested (2017 data). The SA+DTW-system shows a comparable performance to the SA-system

7   Conclusion

The results from last section show two main facts. Firstly, Survival Analysis is suited to be used in batch process monitoring. The information contained in the process variables can be used to model the time-to-end of a batch process, at least for this type of process, with a sufficiently good performance, as judged by some domain experts to which the results were shown. Secondly, the mapping between running and reference batch performed by Dynamic Time Warping is an effective tool to select relevant data from the historical one. The consistent performance between the SA and the SA+DTW system can be interpreted as a good action of the DTW mapping in reducing the noise contained in the data: the relevant information to model the time-to-end is retained when cutting out, on average, more than 99% of the samples available.

These results represent only a first approach to this problem. We think that many improvements could be obtained with careful tuning of the system. Firstly, we used non-optimized parameters for the SA-ML algorithm: the definition of a global performance metric could help in choosing optimal parameters for a given process. Then, the DTW mapping has been performed using the standard version of DTW, but more sophisticated approaches are already present in literature, such as shape-DTW [21]. We think that the performance on the data-selection side could improve given more stable alignment since the version of the algorithm used is susceptible to the noise in the process variables. We think that taking into account this noise is an effective way to stabilize the online alignment and possibly remove the need to aggregate successive predictions of the system to stabilize them.

References

1. Gradient boosting survival analysis. https://scikit-survival.readthedocs.io/en/latest/generated/sksurv.ensemble.GradientBoostingSurvivalAnalysis.html, accessed: 2019-06-14

2. Understanding predictions in survival anlysis. `https://scikit-survival.readthedocs.io/en/latest/understanding_predictions.html`, accessed: 2019-06-14

3. Cox, D.R.: Regression models and life-tables. Journal of the Royal Statistical Society: Series B (Methodological) **34**(2), 187–202 (1972)

4. Gao, X.: On-line monitoring of batch process with multiway pca/ica. In: Principal Component Analysis. IntechOpen (2012)

5. Giorgino, T., et al.: Computing and visualizing dynamic time warping alignments in r: the dtw package. Journal of statistical Software **31**(7), 1–24 (2009)

6. Hothorn, T., Bühlmann, P., Dudoit, S., Molinaro, A., Van Der Laan, M.J.: Survival ensembles. Biostatistics **7**(3), 355–373 (2005)

7. Hunter, J.D.: Matplotlib: A 2d graphics environment. Computing in science & engineering **9**(3), 90–95 (2007)

8. Jones, E., Oliphant, T., Peterson, P., et al.: SciPy: Open source scientific tools for Python (2001–), `http://www.scipy.org/`, [Online; accessed: 2019-06-14]

9. Kassidas, A., Taylor, P.A., MacGregor, J.F.: Off-line diagnosis of deterministic faults in continuous dynamic multivariable processes using speech recognition methods. Journal of Process Control **8**(5), 381 – 393 (1998). https://doi.org/10.1016/S0959-1524(98)00025-0, aDCHEM '97 IFAC Symposium: Advanced Control of Chemical Processes

10. McKinney, W., et al.: Data structures for statistical computing in python. In: Proceedings of the 9th Python in Science Conference. vol. 445, pp. 51–56. Austin, TX (2010)

11. Oliphant, T.E.: A guide to NumPy, vol. 1. Trelgol Publishing USA (2006)

12. Pedregosa, F., Varoquaux, G., Gramfort, A., Michel, V., Thirion, B., Grisel, O., Blondel, M., Prettenhofer, P., Weiss, R., Dubourg, V., et al.: Scikit-learn: Machine learning in python. Journal of machine learning research **12**(Oct), 2825–2830 (2011)

13. Pölsterl, S., Gupta, P., Wang, L., Conjeti, S., Katouzian, A., Navab, N.: Heterogeneous ensembles for predicting survival of metastatic, castrate-resistant prostate cancer patients. F1000Research **5** (2016)

14. Pölsterl, S., Navab, N., Katouzian, A.: Fast training of support vector machines for survival analysis. In: Appice, A., Rodrigues, P.P., Santos Costa, V., Gama, J., Jorge, A., Soares, C. (eds.) Machine Learning and Knowledge Discovery in Databases. pp. 243–259. Springer International Publishing, Cham (2015)

15. Pölsterl, S., Navab, N., Katouzian, A.: An efficient training algorithm for kernel survival support vector machines. arXiv preprint arXiv:1611.07054 (2016)

16. Ramaker, H., van Sprang, E.: Statistical batch process monitoring (01 2004)

17. Ramaker, H.J., van Sprang, E.N., Westerhuis, J.A., Smilde, A.K.: Dynamic time warping of spectroscopic batch data. Analytica Chimica Acta **498**(1), 133 – 153 (2003). https://doi.org/10.1016/j.aca.2003.08.045

18. van Rossum, G.: Python tutorial. Tech. Rep. CS-R9526, Centrum voor Wiskunde en Informatica (CWI), Amsterdam (May 1995)

19. Sakoe, H., Chiba, S.: Dynamic programming algorithm optimization for spoken word recognition. IEEE transactions on acoustics, speech, and signal processing **26**(1), 43–49 (1978)

20. Wang, P., Li, Y., Reddy, C.K.: Machine learning for survival analysis: A survey. arXiv preprint arXiv:1708.04649 (2017)

21. Zhao, J., Itti, L.: Shapedtw: shape dynamic time warping. Pattern Recognition **74**, 171–184 (2018)

Proposal for requirements on industrial AI solutions

Martin W Hoffmann[1], Rainer Drath[2] and Christopher Ganz[3]

[1] ABB AG, Corporate Research, 68526 Ladenburg, Germany
[2] Hochschule Pforzheim, 75175 Pforzheim, Germany
[3] ABB Ltd., 8050 Zürich, Switzerland
`martin.w.hoffmann@de.abb.com`

Abstract. The rise of artificial intelligence (AI) promises productivity gains in industrial practice. While IT technology offers a variety of technological advances, plant owners strive for stability and robustness of the production process. To overcome this tension field, we propose a set of 16 requirements for the development of industrial AI solutions to foster i) the adaptation process, ii) support the solution engineering and iii) ease the embedding into the existing system landscape while respecting iv) safety aspects to build up v) trust into industrial AI solutions. The proposed requirements can guide industrial stakeholders to focus on the right solution approach for specific production challenges and support them in voicing their own needs towards novel AI solutions. This will help AI developers to speed up time-to-market as well as to increase market acceptance of industrial AI solutions. Overall, specifying requirements on industrial AI will foster the acceptance and utilization rates of AI solutions in industrial practice.

Keywords: Artificial intelligence, AI, industrial AI, industrial production, requirements, digital twin, autonomy, manufacturing, Industrie 4.0, Industry 4.0, use-case.

1 Introduction

The introduction of software methods into industrial automation is currently a key source of innovation. E.g., the announced fourth industrial revolution (Industrie 4.0) aims for the introduction of internet technology into production and promises productivity gains across all phases of an industrial plant [1]. While IT technology promises a variety of possibilities or opportunities, plant owners aim for stability and robustness of the plant, requiring reliability and proven technology. This is a field of tension between feasibility and stability and related industrial requirements have been formulated [2].

The same now applies with the introduction of artificial intelligence (AI) into production. Not everything that is possible with mainstream AI is applicable in industry. Therefore, the authors propose a set of industrial requirements for AI solutions as a basic guideline for industrial AI developers and vendors.

© The Author(s) 2021
J. Beyerer et al. (Hrsg.), *Machine Learning for Cyber Physical Systems*, Technologien für die intelligente Automation 13,
https://doi.org/10.1007/978-3-662-62746-4_7

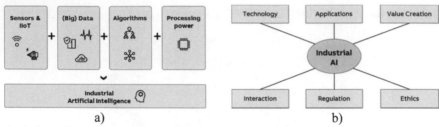

a)   b)

Fig. 1. a) Ingredients comprising Industrial AI. b) Topics of discussion around industrial AI.

1.1   Usage of AI in Industrial Production

The application of AI technologies in industrial is not a new topic and has been subject to scientific investigations also during previous periods of AI research, e.g. [3,4]. More recently, research focused on the application of AI techniques in fault diagnosis and predictive maintenance [5-12] as well as decision support systems [5,13]. Current academic investigations aim at the coordination of the introduction of AI into all layers of production systems [14] as well as production-wide maintenance processes [15].

1.2   Industrial AI

Besides existing definitions of Industrial AI [16,17], we define Industrial Artificial Intelligence for this study as the combination of sensors & IIoT, (big) data, algorithms and processing power (Fig. 1a). However, the discussions about Industrial AI, as the authors experience them in daily practice, are not limited to technological aspects of AI, but rather mix questions about technology, industrial applications, value creation, human-AI-interaction, regulatory aspect as well as ethics (Fig. 1b). These broader understanding of Industrial AI bring up additional requirements from an industrial perspective, which need to be addressed for Industrial AI to unfold its full (positive) impact on an industrial production.

2   Requirements on industrial AI

The presented requirements were collected during the years 2017 and 2018 in an unstructured fashion during several industrial research projects and end-customer discussions in the field of AI-enabled production systems. The collected requirements can be sorted in five categories (Tab 1), which build upon another (Fig. 2).

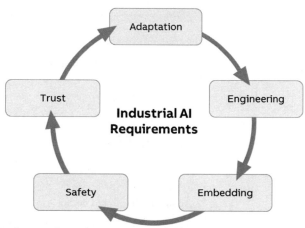

Fig. 2. Interdependence of requirements on Industrial AI.

2.1   Adaption of Industrial AI systems

The introduction of novel technology into production environments brings some specific requirements with it, as we deal mostly with existing production facilities, so-called brownfield environments. Brownfield environments are characterized by a zoo of existing IT and OT systems, as well as experienced employees running the production.

1) Stepwise introduction. Industrial AI should be introduced step by step so that it can initially validate itself in the context of production. A parallel operation with conventional automation technologies is desirable in terms of confidence generation. The power of decision should be gradually transferred from the conventional automation technology to the industrial AI, similar to the gradual increase of autonomy of the system [18]. It is to be noted though that for safety and reliability reasons, some critical functionality may remain in a conventional automation system. The Industrial AI system will thus be operating under the supervision of the safety governing system.

2) Human in the loop / system autonomy. During the introduction phase of Industrial AI in production systems, human should remain in control of all decisions, i.e. the Industrial AI may only serve as expert system or decision support system, corresponding to lower levels of autonomy [18]. Only by time, the autonomy of the Industrial AI may increase.

3) Data availability. AI systems usually require massive amounts of data. In industrial settings we usually struggle far more with the data quality [15], including the amount of relevant information contained in large amounts of machine data [19-21]. Furthermore, data from heterogeneous OT and IT systems needs to be accessed, usually bringing legal, contractual, cultural, commercial, technical and security topics on the table [21].

Table 1. Overview of proposed requirements on Industrial AI solutions.

Area	Requirement
Adaption	1. Stepwise introduction
	2. Human in the loop
	3. Data availability
Engineering	4. Virtual learning
	5. Adaptation
	6. Simplicity (hiding of complexity)
Embedding	7. Stacking of AI decisions
	8. Trust space and borders
	9. Knowledge distribution
Safety / Security	10. Safety
	11. Robust against adversarial inputs
Trust	12. Traceability and transparency of decisions
	13. Bias-free
	14. Confidence measure
	15. Trust / quality classification
	16. Proof of capabilities

2.2   Engineering of Industrial AI systems

Once general problems with the introduction of Industrial AI systems are overcome, the engineering of such novel systems requires special attention.

4) Simplicity / hiding of complexity. Industrial AI solutions may not be designed a way such that the adaptation to new production lines or customer settings requires eminent engineering efforts, as this would counter a main advantage of AI-based solutions. The user of AI-based production systems will most-likely not have a mathematical or computer science education and thus also the usage and operation of industrial AI systems shall be simple and hide all otherwise required technical complexity. A more simplistic design of Industrial AI solution will also lead to greater robustness of the system [23].

5) Virtual learning. Industrial AI should learn over time and the learning should not be limited to physical tests but should additionally be executed in virtual environments as much as possible, e.g. utilizing digital twin concepts [24-26]. In the result, e.g. a robot can optimize its path virtually.

6) Adaptation. Industrial AI should continuously adapt its capabilities to a changing production environment. Environmental changes (e.g. position drift) need to be detected and handled correctly.

2.3   Embedding of Industrial AI system in existing production system landscape

Industrial AI systems independent of their embedding in either greenfield or brownfield environments will interact with several other OT and IT systems in the production facility and beyond. Besides well-known technical requirements of the interfaces between such systems, further requirements will come up in light of the broader capabilities of Industrial AI systems.

7) Trust space and trust borders. In case that industrial AI provides functionality, which cannot be achieved with traditional automation technology (e.g. gripping of loose parts in box) and hence no classic system backup is possible, a dedicated checkpoint is required which allows a human to prove the plausibility of taken decisions.

8) Knowledge distribution. Industrial AI should be able to distribute its knowledge and learnings to other industrial AIs, either directly or via superior systems or communication networks.

9) Stacking of AI decisions. Industrial AI should initially not base its conclusions on data that have themselves been created by another AI.

2.4   Safety and Security of Industrial AI systems

Safety and security are both in the OT and IT world hard requirements, although failures in production systems such as chemical processes or power plants can cause far bigger damage than compromised mainstream AI applications. The individual requirements regarding safety and security may fill complete studies by their own. At this, we would like to only highlight two requirements exemplarily.

10) Safety. Vendors and providers of industrial AI-enabled machines or production systems need to ensure that the they work safe according to the EU Machinery Directive 2006/42/EC [27] as well as categorized according to IEC61508 Safety Integrity Levels (SIL), i.e. the industrial AI may not pose a danger to humans in any possible operation condition and the risk of failure should be made transparent according to the IEC reaching a best possible SIL. Industrial AI solution should in the future also be able to prove that they are safe even under recursive self-improvement [28].

11) Robustness against adversarial inputs. Industrial AI needs to be robust against accidental and intended adversarial inputs to ensure a maximum of protection of the production process.

2.5   Trust in functionality of Industrial AI systems

Production workers and managers fully trust today's production systems. This trust needs to be carried over into AI-enabled production systems, especially in the light of prominent examples of failing mainstream AI [29].

12) Traceability and transparency of decisions. Industrial AI should be able to explain its decisions, e.g. by means of visualizations. Errors in the industrial AI's assumptions shall be recognizable and correctable. Example: if a workpiece is recognized by industrial AI, the considered elements of the workpiece should be visualized. False assumptions should be highlighted and correctable.

13) Trust/quality classification. Industrial AI should be divided into trust/quality classes, which are backed by statistics. This is to express the probability of failure of the AI, e.g. determined by experiments or field tests. Industrial AI can be stacked if the underlying AI fulfills a sufficient trust category.

14) Proof of capability. Industrial AI should allow to check its capabilities and limitations in a determined and safe space, e.g. in a virtual environment or a test run.

15) Bias-free. Industrial AI needs to be constructed free of bias, e.g. treating equipment of all vendors in the same manner. Besides still feeling superior to machines in the consumer world when looking at failing AI [29], Industrial AI may never suffer from bias in data leading to a negative impact on production. As such Industrial AI systems need to be thoroughly tested before the usage in a productive environment, e.g. in the virtual layer of a CPS.

16) Confidence measure. Expectation on Industrial AI system higher than on industrial worker or consumer AI solution. Especially in European countries Industrial AI system are expected to have 100% solution rate and 0% error rate. This of course cannot be achieved, neither by a technical system nor by a human expert. It is therefore required that Industrial AI system provide a confidence measure together with decision making or action recommendations.

3   Discussion

The aim of the presented study is to trigger further discussions about requirements on the introduction of AI into industrial production environments. The presented list of requirements may not be conclusive but covers many aspects specific to Industrial AI solutions. Further requirements left out in the presented list may cover topics such as data / analytics privacy [30], ethical implication [31], potential malicious use of AI [32] as well as AI-vendor lock-in [23].

The term of "Industrial AI" has previously been defined differently to the definition presented in the introduction of this paper [16,17]. Our definition is however not contradicting the previous definitions, but rather sharpens the understanding of Industrial AI in the context of the present study. Previously, the definition of Industrial AI also included application requirements [17] or Industrial AI was thought to "function as a bridge connecting academic research outcomes in AI to industry practitioners" [16], which both fit to the topics surrounding Industrial AI (Fig 1b), from our perspective.

It is interesting to note that some of the presented requirements correspond with identified requirements on Industrial AI correspond to research fields recently identified for Scientific Machine Learning [33]. We foresee future developments of Industrial AI in the areas of i) production system autonomy [18], ii) product life cycle management [15], especially because AI life cycles will become more complex to manage [23], iii) virtual industrial assistants [34], iv) explainable AI [35] and v) the seamless fusion of different data pools in production sites.

4   Conclusion

The introduction of Industrial AI brings up a tension field between plant owners striving for reliability and stability of the production and AI technologies entering the production systems. We proposed a set of 16 requirements in five categories as guidelines for Industrial AI developers as well as Industrial AI users to foster the adaptation of the new opportunities arising from AI in the industrial domain.

Acknowledgements

The work has been supported by the EU ECSEL project Productive 4.0.

References

1. Krueger, MW, et al. "A new era: ABB is working with the leading industry initiatives to help usher in a new industrial revolution." ABB Review 4/2014: 70-75, 2014.
2. Drath, R and Horch, A. "Industrie 4.0 – hit or hype?", IEEE Industrial Electronics Magazine 01/2014; 8(2):56-58, 2014.
3. Fox, M. S. "Industrial applications of artificial intelligence." Robotics 2.4: 301-311, 1986.

4. Parunak, H. "Applications of distributed artificial intelligence in industry." Foundations of distributed artificial intelligence 2, 1996.
5. Yam, RCM, et al. "Intelligent predictive decision support system for condition-based maintenance." The International Journal of Advanced Manufacturing Technology 17.5: 383-391, 2001.
6. Li, Z, Wang, Y and Wang, K-S. "Intelligent predictive maintenance for fault diagnosis and prognosis in machine centers: Industry 4.0 scenario." Advances in Manufacturing 5.4: 377-387, 2017.
7. Sarda-Espinosa, A, Subbiah, S and Bartz-Beielstein, T. "Conditional inference trees for knowledge extraction from motor health condition data." Engineering Applications of Artificial Intelligence 62: 26-37, 2017.
8. Rødseth, H, Schjølberg, P and Marhaug, A. "Deep digital maintenance." Advances in Manufacturing 5.4: 299-310, 2017.
9. Ottewill, JR., and Orkisz, M. "Condition monitoring of gearboxes using synchronously averaged electric motor signals." Mechanical Systems and Signal Processing 38.2: 482-498, 2013.
10. Cecílio, IM., et al. "Nearest neighbors method for detecting transient disturbances in process and electromechanical systems." Journal of Process Control 24.9: 1382-1393, 2014.
11. Amihai, I, et al. "Modeling Machine Health Using Gated Recurrent Units with Entity Embeddings and K-Means Clustering." 2018 IEEE 16th International Conference on Industrial Informatics (INDIN), 2018.
12. Amihai, I, et al. "An Industrial Case Study Using Vibration Data and Machine Learning to Predict Asset Health." IEEE 20th Conference on Business Informatics (CBI). Vol. 1, 2018.
13. Atzmueller, M, et al. "Big data analytics for proactive industrial decision support." atp magazin 58.09: 62-74, 2016.
14. Wan, J, et al. "Artificial intelligence for cloud-assisted smart factory." IEEE Access 6: 55419-55430, 2018.
15. Zhang, Y, et al. "A big data analytics architecture for cleaner manufacturing and maintenance processes of complex products." Journal of Cleaner Production 142: 626-641, 2017.
16. Lee, J, et al. "Industrial Artificial Intelligence for industry 4.0-based manufacturing systems." Manufacturing letters 18: 20-23, 2018.
17. Zhang, , et al. "A reference framework and overall planning of industrial artificial intelligence (I-AI) for new application scenarios." The International Journal of Advanced Manufacturing Technology: 1-23, 2018.
18. Gamer, T. et al., "The Autonomous Industrial Plant – Future of Process Engineering, Operations and Maintenance," in 12th International Conference on Dynamics and Control of Process Systems (DYCOPS), 2019.
19. Gitzel, G, Turrin, S and Maczey, S. "A data quality dashboard for reliability data." 2015 IEEE 17th Conference on Business Informatics. Vol. 1. IEEE, 2015.
20. Gitzel, R. "Data Quality in Time Series Data: An Experience Report." CBI (Industrial Track). 2016.
21. Gitzel, R, Subbiah, S and Ganz, C. "A Data Quality Dashboard for CMMS Data." ICORES. 2018.
22. Deloitte LLP, "Realising the economic potential of machine-generated, nonpersonal data in the EU", available at https://www.vodafone.com/content/dam/vodafone-images/public-policy/reports/pdf/Realising_the_potential_of_IoT_data_report_for_Vodafone.pdf, last accessed 24.04.2019, 2018.
23. Dutta, D, et al. "Towards #consistentAI." First International Conference on Artificial Intelligence for Industries (AI4I), 2018.

24. Tao, F, et al. "Digital twin-driven product design, manufacturing and service with big data." The International Journal of Advanced Manufacturing Technology 94.9-12: 3563-3576, 2018.
25. Wagner, C, et al. "The role of the Industry 4.0 asset administration shell and the digital twin during the life cycle of a plant." 22nd IEEE International Conference on Emerging Technologies and Factory Automation (ETFA), 2017.
26. Malakuti, S, and Grüner, S. "Architectural aspects of digital twins in IIoT systems." Proceedings of the 12th European Conference on Software Architecture: Companion Proceedings, 2018.
27. European Commission, "Directive 2006/42/EC of the European Parliament and of the Council of 17 May 2006 on machinery, and amending Directive 95/16/EC (recast)", Official Journal of the European Union, L 157/24, 2016.
28. Yampolskiy, RV. "Artificial intelligence safety engineering: Why machine ethics is a wrong approach." Philosophy and theory of artificial intelligence. Springer, Berlin, Heidelberg. 389-396, 2013.
29. Yampolskiy, RV and Spellchecker, MS. "Artificial intelligence safety and cybersecurity: A timeline of AI failures." arXiv preprint arXiv:1610.07997, 2016.
30. Marrella, A et al. "Privacy-Preserving Outsourcing of Pattern Mining of Event-Log Data-A Use-Case from Process Industry." IEEE International Conference on Cloud Computing Technology and Science (CloudCom), 2016.
31. EC High Level Expert Group on Artificial Intelligence, "Ethics Guidelines for trustworthy AI", https://ec.europa.eu/digital-single-market/en/news/ethics-guidelines-trustworthy-ai (accessed 03.06.2019), 2019.
32. Brundage, M, et al. "The malicious use of artificial intelligence: Forecasting, prevention, and mitigation." arXiv preprint arXiv:1802.07228, 2018.
33. Baker, N, et al.. "Workshop Report on Basic Research Needs for Scientific Machine Learning: Core Technologies for Artificial Intelligence." United States: N. p., 2019. Web. doi:10.2172/1478744, 2019.
34. Schmidt, B, et al. "Industrial Virtual Assistants: Challenges and Opportunities." Proceedings of the 2018 ACM International Joint Conference and 2018 International Symposium on Pervasive and Ubiquitous Computing and Wearable Computers, 2018.
35. Gunning, D. "Explainable artificial intelligence (xai)." Defense Advanced Research Projects Agency (DARPA), nd Web, 2017.

Information modeling and knowledge extraction for machine learning applications in industrial production systems

Stefan Windmann[1] and Christian Kühnert[2]

[1] Fraunhofer IOSB-INA, Fraunhofer Center for Machine Learning
Campusallee 6, 32657 Lemgo, Germany
`stefan.windmann@iosb-ina.fraunhofer.de`
[2] Fraunhofer IOSB, Fraunhofer Center for Machine Learning
Fraunhoferstraße 1, 76131 Karlsruhe, Germany
`christian.kuehnert@iosb-ina.fraunhofer.de`

Abstract. In this paper, a new information model for machine learning applications is introduced, which allows for a consistent acquisition and semantic annotation of process data, structural information and domain knowledge from industrial productions systems. The proposed information model is based on Industry 4.0 components and IEC 61360 component descriptions. To model sensor data, components of the OGC SensorThings model such as data streams and observations have been incorporated in this approach. Machine learning models can be integrated into the information model in terms of existing model serving frameworks like PMML or Tensorflowgraph. Based on the proposed information model, a tool chain for automatic knowledge extraction is introduced and the automatic classification of unstructured text is investigated as a particular application case for the proposed tool chain.

Keywords: machine learning, information modeling, model serving, knowledge extraction

1 Introduction

Data in industrial production systems is usually stored in a heterogeneous way, using a large variety of data formats and semantics. The integration of these data sources, which cover besides process data also structural information, domain knowledge and process documents, is an essential prerequisite for the successful application of machine learning and optimization methods in the context of industrial production. Therefore, different data sources for machine learning applications (see Fig. 1) need to be fusioned and semantically annotated with structural information and domain knowledge.

J. Beyerer et al. (Hrsg.), *Machine Learning for Cyber Physical Systems*, Technologien für die intelligente Automation 13,
https://doi.org/10.1007/978-3-662-62746-4_8

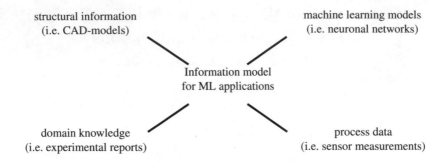

structural information
(i.e. CAD-models)

machine learning models
(i.e. neuronal networks)

Information model
for ML applications

domain knowledge
(i.e. experimental reports)

process data
(i.e. sensor measurements)

Fig. 1. Knowledge needed about a production process when using machine-learning algorithms. Fusioning all information leads to the proposed information model for ML applications.

This paper makes the following contributions to this objective:

– A unified information model for machine learning applications in production systems is proposed.
– Methods for the transformation of plant knowledge into the unified information model are introduced.

The development of appropriate information models for industrial production systems is an essential subject of the current research in the context of Industry 4.0 [4]. Examples are the reference architecture Industry 4.0 (RAMI4.0) [3], modeling languages, which allow for a structured and component-based of production plants, such as AutomationML [9], and industrial communications standards with information modeling capabilities such as OPC-UA [6]. However, such general approaches are usually not tailored to the specific requirements of machine learning approaches. In addition to these approaches, specialized information models exist, e.g. models for the sensor data acquisition like the OGC SensorThings model [1] or standardized XML-descriptions of machine learning models such as PMML [10].

In this paper, an information model for machine learning applications in production environments is proposed, which is build upon general I4.0 components and specialized information models for process data and machine learning models. In addition, a tool chain is introduced, which enables information to be automatically extracted from sensor data and other information sources and to be stored in the form of a corresponding information model. In particular, the extraction of features such as parallel automata and the automatic classification of documents from production environments are considered.

The remaining part of this paper is structured as follows: The proposed information model for machine learning applications is introduced in section 2. The tool chain for knowledge extraction is described in detail in section 3. Finally, a conclusion is given in section 4.

2   Information modeling

The essential purpose of the proposed information model is to describe production plants in such a way that machine learning applications can be straightforwardly applied to them. Figure 2 shows the complete entity relationship diagram of the proposed information model. The information model is build upon Industry 4.0 components [5]. Following a hierarchical order, components can be arranged in a tree form using subcomponents, which are specified as attributes of the entity *Component*. The information required to describe a production plant

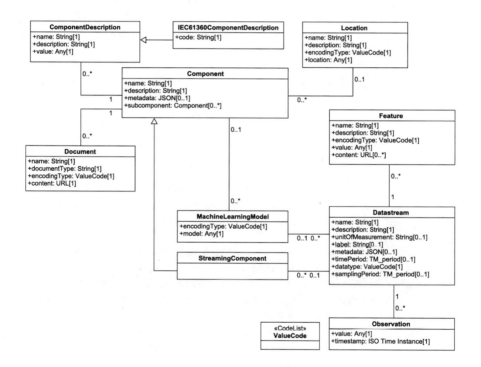

Fig. 2. Proposed information model

for machine learning applications can be roughly divided into four classes, which determine the structure of the proposed information model:

- **Structural information:** Machine learning approaches require information about the construction of the production plant, e.g. about the positions of sensors and actuators. Structural information of the production plant, which comes e.g. from CAD models, is mainly stored in the component tree using the entity *Component*. To store geospatial data for the individual components, the entity *Location* has been adopted from the OGC SensorThings model [1].

- **Domain knowledge:** In a production process, lots of domain knowledge is is available, which can be exploited for machine learning. This covers e.g. component descriptions, experimental reports, log-files, thresholds for sensor signals or information about maintenance cycles. The incorporation of such domain knowledge is mainly implemented by means of the entities *ComponentDescription*, *Feature* and *Document*. Information about the individual components (i.e. sensor replaced, sensor cleaned, etc.) is stored in the entity *ComponentDescription*, where it is possible to add e.g. the Industry 4.0 admin shell [5] and the IEC 61360 component descriptions [2] but also domain specific component descriptions. The entity *Feature* is used to add information to a *Datastream* (i.e. by adding labels, standard deviations, mean values etc.). Text documents are integrated into the proposed information model using the entity *Document*. In doing so, either links to the particular documents are used or the documents are made machine-readable by following the approach described in section 3.
- **Process data:** The process data, which is acquired and fusioned from different data sources, forms the basis for most of the existing machine learning approaches for production plants. The information model has to provide detailed information on the type and the usage of the measurements. For this purpose, the concept of data streams has been taken from the OGC SensorThing model [1]. The entity *StreamingComponent* is used to model components containing a *Datastream*. A *Datastream* describes in detail what and how the component is measuring (i.e. the unit, feature of interest, metadata). Hereby, it is worth noting that a streaming component can also be an experimental protocol, a log-file or even the operator controlling the process. To store the individual measurements, the entity *Observation* has been incorporated into the information model, which is connected using a one to many relationship to *Datastream*.
- **Machine learning models:** When using machine learning algorithms in a production environment those algorithms need to be served correctly. This serving can be in its simplest form the description of a numerical processing pipeline or in a more complex way a graph, which is stored in an XML file. Machine learning models are incorporated into the information model by using the entity *MachineLearningModel*. The proposed information model for ML applications has the possibility to store those models in a standardized way using XML descriptions such as the predictive markup model language (PMML) [10], Tensorflow or Pytorch, but it is also possible to store the served model in terms in a tailor-made, non-standardized format.

The information model covers in summary 10 entities. Details on the particular entities are described in the appendix.

3   Tool chain for knowledge extraction

Machine learning methods used to optimize industrial production systems usually require the evaluation of huge and diverse data sets for successful operation. In such application cases, the manual extraction of information is time consuming and error-prone. Hence, a tool chain has been developed in the present work, which allows for automatic knowledge extraction using the proposed information model (see Fig. 2).

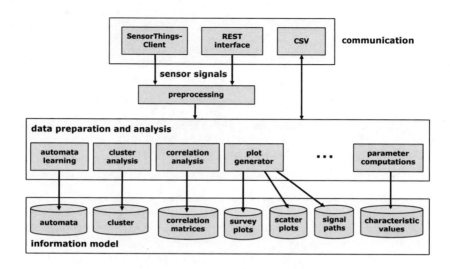

Fig. 3. Tool chain for knowledge extraction

The core of the tool chain consists of distinct ML and statistic methods, which accomplish automatic data extraction (e.g. the extraction of machine learning models such as automata [11], or features such as correlation matrices, data clusters, characteristic values, surveys, or scatter plots). Besides, different interfaces for data acquisition are provided, particularly a client for the OGC SensorThingsAPI, a REST interface and an interface for the offline import of CSV files. The extracted knowledge is integrated in the proposed information model, particularly in the entities *MachineLearningModel* and *Feature*, and can be used for plant optimization.

The proposed tool chain is also used for the automatic classification of unstructured text. In this use case, a data pool with several types of text documents is available, which contains e.g. operation manuals, shift books and repair instructions. In particular, documents from nine classes are available: operation manuals (MNL), glossaries (GLS), tables of contents (CNT), labels (LBL), security instructions (SEC), indexes (IDX), parts lists (PRT), technical data (DAT) and service notes (SRV). A document classification is used to assign an appropriate document type to each document to allow for structured access to the particular document. Based on the document type, the documents are automatically inserted into the *Document* entity of the proposed information model.

Initial evaluations of the document classification have been conducted in [12]. In doing so, K-Nearest-Neighbor classification [7] and Bayesian classification [8] have been investigated for documents of the SmartFactoryOWL, an evaluation platform for cyber-physical production systems (see Fig. 4). Altogether, each classifier has been learned

Fig. 4. SmartFactoryOWL

from 194 documents. Evaluations have been conducted on 66 documents. In this process, 65 of the 66 documents have been correctly classified for the k-NN classifier. For Naive Bayes, only 38 documents have been correctly classified. Furthermore, the k-NN classifier has been observed to significantly outperform the Naive Bayes classifier with respect to computational complexity. The runtime of the k-NN classifier on 66 documents amounts to 11s, while the Naive Bayes classifier requires 70s for the same task. Altogether, the k-NN classifier has been shown to be more suitable for the investigated application case.

4   Conclusion

In this article, a new information model was introduced, which makes it possible to record and present in a structured way the information relevant for the use of machine learning methods in production environments. Furthermore, it was shown that parts of the modeled information such as process models, features or document types can be automatically extracted from the available data. In a next step it is planned to instantiate the information model as well as the tool chain at a glass bending plant and at the SmartFactoryOWL, a demonstration and evaluation platform for cyber-physical production systems.

5   Acknowledgement

This work was partly developed in the Fraunhofer Cluster of Excellence "Cognitive Internet Technologies".

Appendix: Entities of the proposed information model

In this section, the entities of the proposed information model (Figure 2) are described in detail:

- **Component:** In the information model, the entity *Component* represents an object that can either be a single part or a composition of several other components (i.e. tubes, motors, valves). A component can therefore be seen as a part of the overall system. The attributes are (1) name: meaningful name and (2) description: detailed description of the component; (3) metadata: additional information about the component.

- **Location:** The entity *Location* describes the position of the component. This can be a geo position, a postal address or a position of a component inside the production system or somewhere else. The attributes are (1) name and (2) description: meaningful name and description of the location; (3) encodingType: value code of the location, i.e. GeoJSON (application/vnd.geo.json) oder plain text (text/plain) (4) location: value of the location depending on value code.

- **Document:** This entity contains information about documents for a component. The attributes are (1) name: meaningful name of the document; (2) documentType: type of the document, e.g. hand book, service entry or part list (3) encodingType: type of the encoding, e.g. PDF, and (4) content: storage path of the document.

- **ComponentDescription:** The component description contains information about the component, e.g. units, permitted values or schemes for the data transfer in streaming components. The attributes are (1) name, (2) description and (3) value, giving a meaningful name and description to the corresponding value.

- **IEC61360ComponentDescription:** In this entity, component descriptions according to the IEC 61360 norm [2] can be stored. For this, the entity inherits from *ComponentDescription* and contains an additional code, which is used to identify the corresponding IEC 61360 class. Further attributes can be added to the entity according to the IEC 61360 norm.

- **StreamingComponent:** The entity *StreamingComponent* inherits from *Component* and additionally contains the entity *Datastream*. The entity is in particular used to model streaming components like sensors and actors, but it is also suitable for the modeling of users or operators of the production processes.

- **Feature:** This entity is used for a data stream or a streaming component to add further information like a mean value of a data stream, quartile distances or information about the underlying probability distributions. The attributes are (1) name and (2) description: meaningful and description of the feature; (3) encodingType: value code of the feature like double for thresholds, or matrix of double values for histograms; (4) value: value of the feature depending of the defined encodingType and (5) content: path for a more detailed description of the feature.

- **Datastream:** This entity contains data streams, which are captured by a streaming component. The attributes are (1) name and (2) description: meaningful name and description of the data stream; (3) unitOfMeasurement: physical unit of the values being in the *Observation* entity; (4) label: type of the data stream, e.g. faulty or error-free data (5) metadata: additional metadata containing further information about the data stream (6) timeperiod: Time interval in which the observations take place; (7) datatype: datatype of the observations depending on the value code (i.e. double, boolean); (8) samplingPeriod: Used sampling period for the Observations.

– **Observation:** The entity *Observation* contains the process data with the two attributes (1) value: the current value and (2) timestamp: the timestamp of the measurement.
– **MachineLearningModel:** This entity integrates machine learning models. The attributes are (1) encodingType: the encoding determines how the model is stored according to a defined value code (i.e. Predictive Model Markup Language "PMML") and (2) model: the stored model.

References

1. OGC SensorThings API Part 1: Sensing; http://docs.opengeospatial.org/is/15-078r6/15-078r6.html (2016)
2. IEC International Standard 61360: Standard data element types with associated classification scheme (2017)
3. Adolphs, Bedenbender, et al., D.: Referenzarchitekturmodell Industrie 4.0 (RAMI 4.0). Tech. rep., VDI, ZVEI (2015)
4. Bangemann, Bauer, et al., B.: Industrie 4.0 - Technical Assets Grundlegende Begriffe, Konzepte, Lebenszyklen und Verwaltung. Tech. rep., VDI (2015)
5. Bundeswirtschaftsministerium für Wirtschaft und Energie (BMWi): Struktur der Verwaltungsschale - Fortentwicklung des Referenzmodells fuer die Industrie 4.0-Komponente, vol. 1. Spreedruck (2016)
6. Mahnke, W., Leitner, S., Damm, M.: OPC Unified Architecture. Springer Publishing Company, Incorporated, 1 edn. (2009)
7. Runkler, T.A.: Data Analytics Models and Algorithms for Intelligent Data Analysis. Springer Vieweg (2012)
8. Russel, S., Norvig, P.: Artificial Intelligence: A Modern Approach. Prentice Hall (2003)
9. Schleipen, M., Schick, K., Hövelmeyer, T., Okon, M., Wei, J.: Leitfaden "Interoperable semantische Datenfusion zur automatisierten Bereitstellung von sichtenbasierten Prozessführungsbildern (IDA)". Fraunhofer Verlag (2011)
10. The Data Mining Group: PMML: http://dmg.org/pmml/v4-3 (2019), PMML: http://dmg.org/pmml/v4-3
11. Windmann, S., Lang, D., Niggemann, O.: Learning parallel automata of PLCs. In: International Conference on Emerging Technologies and Factory Automation (ETFA) (2017)
12. Windmann, S., Niggemann, O.: Information Retrieval in Industrial Production Environments. In: International Conference on Emerging Technologies and Factory Automation (ETFA) (2018)

Explanation Framework for Intrusion Detection

Nadia Burkart[1], Maximilian Franz[1], and Marco F. Huber[2,3]

[1] Fraunhofer IOSB, Karlsruhe, Germany
[2] Institute of Industrial Manufacturing and Management IFF, University of Stuttgart, Germany
[3] Center for Cyber Cognitive Intelligence (CCI), Fraunhofer IPA, Stuttgart, Germany

Abstract. Machine learning and deep learning are widely used in various applications to assist or even replace human reasoning. For instance, a machine learning based *intrusion detection system (IDS)* monitors a network for malicious activity or specific policy violations. We propose that IDSs should attach a sufficiently understandable report to each alert to allow the operator to review them more efficiently. This work aims at complementing an IDS by means of a framework to create explanations. The explanations support the human operator in understanding alerts and reveal potential false positives. The focus lies on counterfactual instances and explanations based on locally faithful decision-boundaries.

Keywords: Intrusion Detection · Explainable Machine Learning · Counterfactual Explanations

1 Introduction

Advances in machine learning models are associated with an increased complexity of the models. These models appear to end users and even to their developers as black boxes. The reasoning behind the model is often opaque. The research field of explainable machine learning focuses on making models more accessible, transparent and comprehensible for users. Over the past years, there was a surge in approaches for better explainability of the models. Explainable approaches are especially sought after in critical use cases like network-security, medicine or finance. By enabling a lay system user to understand and reproduce the fundamental workings of a machine learning model, trust can be built and improved. In an IDS, explanations of the underlying model can help a system operator to easily understand the model's judgment and reveal potential false positives. In a binary classification task (e.g., classifying suspicious vs. normal behaviour), the concept of a counterfactual explanation is particularly helpful for the human operator as it formalizes a common thought process: *"Why did X happen and not Y?"*. Counterfactual questions are a tool to expose flaws in the underlying decision process. By revealing counterfactuals to the system operator, this could clarify his mental model of a black box classifier and uncover flaws in the model's judgment. We focus on three aspects:

- *Understandability*: Explaining the classification of an instance, based on some form of feature importance.

J. Beyerer et al. (Hrsg.), *Machine Learning for Cyber Physical Systems*, Technologien für die intelligente Automation 13,
https://doi.org/10.1007/978-3-662-62746-4_9

- *Actionability*: Giving practical advice how to change the classification towards the desired result.
- *Simulatability*: Outlining the decision process to allow a user to simulate the behaviour of the model.

In the following, we first give some background of existing work and introduce notations. In Section 3, we then generalize existing counterfactual approaches into the five phases we consider essential for every counterfactual explanation. We slightly adapt modules of existing work, which we evaluate on the IDS scenario.

2   Explanations for Intrusion Detection

We denote by $f : \mathcal{X} \to [0,1]$ a binary black-box classifier that we want to explain. We assume that f is pre-trained as part of an IDS. Hence, f maps so-called attack-vectors $\vec{x}$ from a multidimensional feature-space $\mathcal{X} \subseteq \mathbb{R}^n$ onto the probability that they are malicious instances.

2.1   Surrogate Models

Surrogate models approximate black-box models either locally or globally in an interpretable fashion. One of the best known methods to locally explain black box models by training a surrogate is local model-agnostic explanations (LIME) [1]. Since their work has been thoroughly explained, tested and used [2], we will not elaborate on the specifics of the method. It suffices to note that the idea of LIME is to train a surrogate model g that approximates the original black box classifier f, $g \sim f$, based on training data sampled in a neighborhood around the instance of interest, $\vec{x}_0$. LIME provides a set of feature attributions (see Section 4) derived from the weights of the linear classifier g trained on the sampled data set. These attributions tell the user, which features contributed most significantly to the result.

2.2   Counterfactual Explanation

Laugel et al. [3] note, there is another approach to the local explanation problem, which yields a slightly different interpretation. Namely, what we propose to call *decision boundary centered explanations*. While LIME illustrates which features contribute to an instance, *Local Adversarial Detection (LAD)* [4] and *Local Surrogate* [3] yield a feature attribution that is relevant at a local decision boundary. To do so, it is required to find the decision boundary first and then to train a surrogate on instances located around the decision boundary. Laugel et al. find the decision boundary through random spherical sampling around the instance $\vec{x}_0$. Wachter et al. [5] introduced another solution based on counterfactuals. A counterfactual of $\vec{x}$ is an instance $\vec{x}'$, that yields the opposite classification. Thus, given $\vec{x}$ and f we are searching for $\vec{x}'$ such that $\hat{f}(\vec{x}) \neq \hat{f}(\vec{x}')$, where $\hat{f} : \mathcal{X} \to \{0,1\}, \hat{f}(\vec{x}) \mapsto \lfloor f(\vec{x}) \rceil$, is the binary classifier that yields the predicted class. Ideally, $\vec{x}'$ is *close* to $\vec{x}$ in the feature space $\mathcal{X}$, with respect to some distance metric $d(\cdot, \cdot)$. This formalizes the intuition that the counterfactual should be similar to the original instance. The major contribution from Wachter et al. is to consider the search for a counterfactual as an optimization problem. Formally, Watcher et al. propose to minimize a function

$$L(\vec{x}, \vec{x}', y', \lambda) = \lambda \cdot (f(\vec{x}') - y')^2 + [(1 - \lambda) \cdot d(\vec{x}, \vec{x}')] \times \mathbb{I}(\vec{x}') , \qquad (1)$$

$$\text{where } \mathbb{I}(\vec{x}') = \begin{cases} 0, & \hat{f}(\vec{x}') \neq y' \\ 1, & \hat{f}(\vec{x}') = y' \end{cases}$$

with respect to $\vec{x}'$. With $\lambda \in [0, 1]$, we control the effect of locality, $y' = 1 - y$ denotes the opposite class of the classifier and $\mathbb{I}$ is an optional indicator function. Since the classifier f is a black box, one has to optimize for $\vec{x}'$ using *derivative free* methods (e.g., Nelder-Mead). We elaborate on the methods in Section 3.1. In the following, we are concerned especially with the *decision boundary centered explanations* as they tend to yield more decisive results. We will see that counterfactuals are in fact a by-product of the search for the decision boundary.

3   The Modular Phases of Explanations

We dissect the method of finding *decision boundary centered explanations* into five distinct phases, containing the search for counterfactuals. Also, we present existing approaches for the single phases to give a better intuition (see Figure 1 and Table 1). We start with a given instance $\vec{x}_0$ of class A, an attack instance, of which we want to explain the classification $\hat{f}(\vec{x}_0)$. The goal is to explain why f decided $\vec{x}_0$ to be class A rather than B, a benign instance. This is the specific setting of an IDS described above. The semantic goal of the explanation is to allow the user to judge whether the decision was correct. A consideration that we wanted to keep in mind during all phases is that inference of the model f, or $\hat{f}$ for that matter, might be very expensive. Thus, we aim to keep the number of queries to the black-box small.

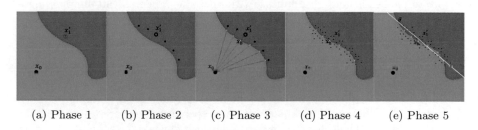

(a) Phase 1   (b) Phase 2   (c) Phase 3   (d) Phase 4   (e) Phase 5

Fig. 1: The five phases of explanations.

Table 1: Overview of the various approaches for the phases 1 to 5, see Section 3.1-3.5

Phase 1	Phase 2	Phase 3	Phase 4	Phase 5
Derivate-Free [5], Growing Spheres [4], Random Sampling [3]	Magnetic Sampling, Random Sampling	Linear Search, Binary Search	Train on sample set, Train on boundary touchpoints	Explanation with interpretable model (e.g., small decision tree)

3.1   Phase 1: Finding the First Counterfactual

The first support point $\vec{x}_1'$, i.e., the first counterfactual, is an instance such that $\hat{f}(\vec{x}_0) \neq \hat{f}(\vec{x}_1')$. As mentioned in Section 2.2, this can be formulated as an optimization problem. Alternatively, we can use random approaches similar to [1] or [4]. Randomly sampling instances in a neighbourhood of the instance $\vec{x}_0$ can be very expensive as the counterfactual might be far away in the feature space of possible instances. Therefore, we use the optimization approach introduced in [5] with minor adaptions. Particularly, we use the distance metric

$$d_m(\vec{x}, \vec{x}') = \frac{1}{n} \sum_i^n \frac{|x_i - x_i'|}{\mathrm{MAD}_i} \ ,$$

that is robust to outliers. Here, n is the dimension of $\mathcal{X}$, x_i denotes the i-th feature value of instance $\vec{x}$ and MAD_i is the median absolute deviation of feature i in the training dataset P according to

$$\mathrm{MAD}_i = \mathrm{median}_{\vec{x} \in P} \left(x_i - \bar{x}_i \right) \ ,$$

with $\bar{x}_i = \mathrm{median}_{\vec{x} \in P}(x_i)$. We normalize over the number of dimensions as our framework aims to be agnostic. Next, (1) retrieves the counterfactuals through

$$\phi(\vec{x}, y', \lambda) = \operatorname*{argmin}_{\vec{x}'} \quad \lambda \cdot (f(\vec{x}') - y')^2 + d_m(\vec{x}, \vec{x}') \ . \tag{2}$$

In our implementation, we minimize (2) with the Nelder-Mead simplex algorithm [6], which is a derivative free method. The result of (2) is the first counterfactual.

3.2   Phase 2: Finding Support Points

Given the first counterfactual $\vec{x}_1' \in \mathcal{X}$, we want to find a set S of instances, such that all $\vec{x}_i' \in S$ are counterfactuals. Literally speaking, they are located on the "opposite side" of the decision boundary. The desired goal is to expand S in order to get a good representation of the actual area where f classifies instances as class B. The idea behind our approach named *MagneticSampling* is to expand the area stepwise into all directions across the dimensions starting from the initial sample $\vec{x}_1'$ until the newly sampled instances are no longer classified as B. For this purpose, we first determine the direction vector $\vec{d} := \vec{x}_1' - \vec{x}_0$ between the original instance and the first counterfactual. We deterministically sample support points $\vec{x}_i'$, $i > 1$, by rotating $\vec{d}$ around $\vec{x}_0$, i.e., taking points with distance $\|\vec{d}\|$ from $\vec{x}_0$ that are in the vicinity of $\vec{x}_1'$, with a fixed discretization step size. This corresponds to taking the support points from the set

$$B(\vec{x}_0, \vec{x}_1', a) = \{\vec{z} \in \mathcal{X} : \|\vec{z} - \vec{x}_0\| = \|\vec{x}_1' - \vec{x}_0\| \text{ and } \|\vec{z} - \vec{x}_1'\| \le a \text{ and } f(\vec{z}) = y'\} \ ,$$

with $\|.\|$ being the Euclidean norm.

Considering only instances around $\vec{x}_1'$ ensures that we find one connected decision boundary and not multiple patches. While possibly neglecting other possible boundaries, this simplifies the explanation [7].

3.3   Phase 3: Finding Decision Boundary

Given the set S of support points or counterfactual points, we approximately locate the decision boundary, which is somewhere on the line segment between $\vec{x}_0$ and any $\vec{x}_i' \in S$. We denote the segment by $L_i(v) := \vec{x}_0 + v \cdot (\vec{x}_i' - \vec{x}_0)$ with $v \in [0,1]$. The result of this phase is some abstract representation of the possibly sophisticated decision boundary in local proximity to $\vec{x}_0$. To give an intuition, this could mean a set of points B such that each $\vec{x}_b \in B$ is on the decision boundary (a border touchpoint) [3], or it could be a polygon enclosing the decision boundary in a given segment. Considering the way we sampled our support points, we can assume that the value of $\hat{f}$ develops monotonously on the segment L_i. Note that this does not have to be true for the prediction probability f. Given this assumption we can use binary search on the segments to approximately locate $\vec{x}_b = L_i(v)$ for some v and thereby reduce the number of queries to our black-box f from $\mathcal{O}(n)$ to $\mathcal{O}(\log(n))$.

3.4   Phase 4: Train Explainer on Sample Set

Using the representation of the local decision boundary from Phase 3 we sample a set T of instances around the decision boundary. Given T we train a simple model g, called surrogate, to approximate the decision boundary locally. Similar to [1], we constrain the complexity $\Omega(g)$ of the model by imposing constraints like maximum depth for decision trees or number of non-zero weights for linear classifiers. Formally, we obtain g out of a class of models G (e.g. decision trees, linear models, ...) through

$$\varphi(T, f, L') = \operatorname*{argmin}_{g \in G} \sum_{\vec{x} \in T} L'\big(f(\vec{x}), g(\vec{x})\big) \ ,$$

where L' is some loss function (e.g. Mean-Squared-Error loss).

The framework allows manually or automatically limiting the number of features considered by the surrogate g. If no previous knowledge is available to select features, Least-Angular Regression (LARS, [8]) can be used to determine a restricted feature set.

3.5   Phase 5: Present Explanation & Give *Advice*

Given the results of the previous phases we can now present various explanations. The three major examples are

- *Feature Importance*: As Ribeiro et al. [1] verified, feature importance or attribution, can be a useful way to understand a decision post-hoc.
- *Relative Differences*: We use counterfactual instances revealed in phase one to provide actionable explanations for a user in form of relative differences. See Sec. 4 for an example.
- *Surrogate Visualization*: For the aspect of *simulability*, it is desirable to show a representation of the model to the user. Due to their computational simplicity, decision trees are favorable for this task.

4   Experiment

In this chapter the fidelity of the surrogates and their configuration is evaluated on different data sets . Furthermore, we exemplary present possible explanations for the use case of an IDS. For the IDS2017 [9] and the KDD[10] we trained the MLP classifier on the subset of Web and DOS attacks. The fidelity quantifies how well the surrogate model mimics the behavior of the MLP. Fidelity is the percentage of test examples on which the prediction made by the surrogate matches with the prediction of the trained black box (MLP) [11].

The results for the different configuration by using 10-fold cross-validation are displayed in Table 2 and Table 3. Looking at the results from Table 2 for the IDS data set, we observe that the tree surrogate proposed by the framework consistently outperforms linear approximations trained in LIME fashion and according to our linear approach explained in Section 3. As shown in [12], decision trees also far better in terms of human interpretability. In short, the decision tree trained on the decision boundary (DB-tree) is both more accurate and more interpretable. For the random configuration illustrated in Table 3 mostly LIME outperforms DB-Linear and DB-tree. The results of Table 2 and 3 illustrate that the systematic approach (Nelder Mead/Magnetic Sampling) is more effective than LIME and the random approaches.

We continue with a visualization of the possible explanations of our framework, but limit ourselves to the rather novel approaches of *relative difference*

Table 2: Fidelity for Nelder Mead/Magnetic Sampling

Data set	LIME	DB-Linear	DB-Tree
IDS [9]	0.85	0.87	**0.97**
KDD [10]	0.93	0.96	**0.99**
Heloc [13]	0.86	0.96	**0.97**
Credit [14]	0.95	**0.99**	**0.99**

Table 3: Fidelity for Random

Data set	LIME	DB-Linear	DB-Tree
IDS [9]	0.84	**0.91**	0.85
KDD [10]	**0.92**	0.84	0.83
Heloc [13]	**0.92**	0.78	0.82
Credit [14]	**0.96**	0.87	0.92

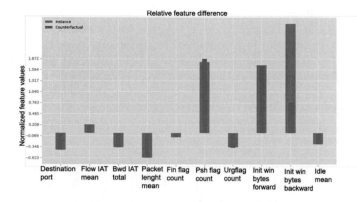

Fig. 2: Relative feature difference between instance and counterfactual (Data set: IDS)

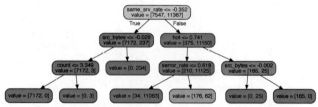

Fig. 3: Decision Tree trained on the decision boundary (Accuracy 0.99) (Data set: KDD)

and *surrogate visualization* for brevity. The feature attribution we can retrieve, matches in its nature that of LIME and can help a user to *understand* a decision. The *Relative Difference* method on the other hand, makes use of the counterfactual to give actionable advice. Figure 2 shows the differences between the instance and its counterfactual for the ten most significant features. It quickly reveals that the high value of *Init win bytes backward* caused the erroneous classification as an attack.

Surrogate visualization on the other hand helps the user to *simulate* the decision process. For this task, the decision tree depicted in Fig. 3 is suited best, as the effort for manually inferring a prediction is low [12].

5   Summary

In this paper, a theoretical framework for modular decision boundary focused explanations was proposed. By distributing the training of an explainable surrogate in different modules, flexibility and variety is introduced. The aspect of generating decision boundary centered explanations allows easily generating counterfactuals. Due to the increasing demand for explainable machine learning systems, various approaches should be pursued in parallel. With this work we contribute to the field of model-agnostic analysis, for which many methods have been proposed before [15]. Depending on the requirements of the application, other approaches like those proposed by Pearl et al. [16] ought to be pursued in parallel. By reviewing the literature on explainable machine learning, we have encountered a confusing ambiguity when it comes to terminology. Clear research directions and notation ought to be introduced. More user studies like [12] are needed to gain more insights of how understanding and actionability really can be obtained.

References

1. M. Tulio Ribeiro, S. Singh, and C. Guestrin, ""Why Should I Trust You?": Explaining the Predictions of Any Classifier," *ArXiv e-prints*, Feb. 2016.
2. S. Mishra, B. L. Sturm, and S. Dixon, "Local interpretable model-agnostic explanations for music content analysis.," in *ISMIR*, pp. 537–543, 2017.
3. T. Laugel, X. Renard, M.-J. Lesot, C. Marsala, and M. Detyniecki, "Defining Locality for Surrogates in Post-hoc Interpretablity," *ArXiv e-prints*, June 2018.
4. X. Renard, T. Laugel, M.-J. Lesot, C. Marsala, and M. Detyniecki, "Detecting Potential Local Adversarial Examples for Human-Interpretable Defense," *ArXiv e-prints*, Sept. 2018.
5. S. Wachter, B. D. Mittelstadt, and C. Russell, "Counterfactual explanations without opening the black box: Automated decisions and the GDPR," *CoRR*, vol. abs/1711.00399, 2017.
6. J. A. Nelder and R. Mead, "A simplex method for function minimization," *The computer journal*, vol. 7, no. 4, pp. 308–313, 1965.
7. C. Molnar, "Interpretable machine learning," *A Guide for Making Black Box Models Explainable*, 2018.
8. B. Efron, T. Hastie, I. Johnstone, R. Tibshirani, *et al.*, "Least angle regression," *The Annals of statistics*, vol. 32, no. 2, pp. 407–499, 2004.
9. I. Sharafaldin, A. H. Lashkari, and A. A. Ghorbani, "Toward generating a new intrusion detection dataset and intrusion traffic characterization.," in *ICISSP*, pp. 108–116, 2018.
10. M. Tavallaee, E. Bagheri, W. Lu, and A. A. Ghorbani, "A detailed analysis of the kdd cup 99 data set," in *2009 IEEE Symposium on Computational Intelligence for Security and Defense Applications*, pp. 1–6, IEEE, 2009.
11. M. Craven and J. W. Shavlik, "Extracting tree-structured representations of trained networks," in *Advances in neural information processing systems*, pp. 24–30, 1996.
12. S. A. Friedler, C. D. Roy, C. Scheidegger, and D. Slack, "Assessing the local interpretability of machine learning models," *arXiv preprint arXiv:1902.03501*, 2019.
13. "Heloc explainable ml challenge." https://community.fico.com/s/explainable-machine-learning-challenge. Accessed: 2019-03-01.
14. H. Hofmann, "Statlog data set." https://archive.ics.uci.edu/ml/datasets/statlog. Accessed: 2019-06-13.
15. A. Shrikumar, P. Greenside, and A. Kundaje, "Learning important features through propagating activation differences," *arXiv:1704.02685*, 2017.
16. J. Pearl *et al.*, "Causal inference in statistics: An overview," *Statistics surveys*, vol. 3, pp. 96–146, 2009.

Automatic Generation of Improvement Suggestions for Legacy, PLC Controlled Manufacturing Equipment Utilizing Machine Learning

Wolfgang Koehler [0000−0002−8046−3501] (✉) and Yanguo Jing [0000−0001−9581−4215]

School of Computing, Electronics and Mathematics, Faculty of
Engineering, Environment and Computing, Coventry University
Priory Street, Coventry, UK, CV1 5FB
koehlerw@coventry.ac.uk

Abstract. The manufacturing industry and, for this research, the automotive manufacturing industry specifically, is always on the lookout for opportunities to improve production throughput with a minimum of investment. Identifying these opportunities often requires the observation of the current production process by experts. This paper is the continuation of the previous work 'Automated, Nomenclature Based Data Point Selection for Industrial Event Log Generation'. One of its aims is to provide strategies that can be used to pre-process an in-depth, slightly flawed industrial equipment log to allow for further analysis. The pre-processing is achieved by identifying the flaws, removing the non-value added events and a heuristic methodology to cluster the log into individual sequences. Expert knowledge then is encoded into engineering features to extend the log matrix and prepare it for machine learning model generation for identification of the complete cases. To derive value from the available data, the sequences are plotted into Gantt charts, and eight hypotheses are introduced that allow for automated annotations within this chart to highlight potential areas of improvement. Application of the framework to real life logs, obtained from stations considered bottlenecks within the evaluated automotive body shop, lead to the discovery of improvement potential between two and twelve seconds per cycle.

Keywords: Industrial Logs · Process Mining · Case Clustering

1 Introduction

This research aims to devise an automated framework that will, provided with the code of the programmable logic controller (PLC), monitor the desired production equipment and generate a Gantt chart of its actual sequence while highlighting areas of improvement. The proposed framework has been structured into three distinct approaches.

J. Beyerer et al. (Hrsg.), *Machine Learning for Cyber Physical Systems*, Technologien für die intelligente Automation 13,
https://doi.org/10.1007/978-3-662-62746-4_10

The first function required is automated, nomenclature based data point selection and equipment log generation, as described in detail in the authors' previous publication [1]. Here the goal is the collection of start and end timestamps for all motions within a production cell. The relevant tags to be monitored are, based on their nomenclature, extracted from the PLC program and stored within a SQL database. The monitoring is done with a centralized workstation utilizing an Open Platform Communication (OPC) server. The necessary OPC groups and items are automatically generated. Changes within the status of the PLC tags will trigger an event which logs those changes in the database. In order to evaluate the quality of the obtained data, a quality matrix was devised and applied. The evaluation showed that the records' completeness was above 96% for real-life equipment data.

The second step is machine learning based pre-processing. The obstacle to overcome is clustering the event log data into cases as the raw data do not contain a reliable case identifier. Case clustering was achieved with the part present status within the station and a heuristic approach that allows for the identification of case-related setup, load/unload and reset events. Next, five hypotheses were formulated to create additional features for the data set based on expert knowledge. After tagging the trace classes manually, six different machine learning algorithms were applied with cross-validation. More details can be found in paragraph 3.1.

In chapter 3.2, an expert knowledge-based, heuristic generation of improvement suggestions is introduced. The eight hypotheses postulated were implemented using Python and applied to event logs of four real-life framing re-spot stations. A sequence chart for every style, based on the pre-processed event log was plotted. The issues found were automatically annotated within the same chart, and the findings for the four stations summed up.

2   Related Works

Plant floor systems, as described by Lee [2], were the first step towards the autonomous observation of manufacturing processes. They are logging critical parameters of the process which are used to create KPI (key points of interest) charts and to highlight potential bottlenecks. Next cyber-physical systems started to emerge. Their goal, to create a digital clone of the real-life production equipment, which can be used to create simulations and derive predictions, was also documented by Lee [3]. Jaber et al. [4] showed that predictions regarding required maintenance could also be obtained by applying machine learning techniques to vibration sensor data. The results could help to move the time of preventive maintenance closer to the predicted time of failure thus realising additional savings. Banerjee et al. [5] propose a similar approach. Instead of using vibration sensors, which normally are not an integral part of manufacturing equipment, they are utilising the already available sensors for fault detection.

Processes can not only be found in manufacturing but also for business transactions. Van der Aalst [6] started at the beginning of this century the development of the research field of Process Mining. The aim is to discover the underlying process model of such business transactions based on logged transaction data. Hu et al. [7] realised that the proposed algorithms might also be beneficial for the discovery of process models within 'flexible manufacturing systems'. They proclaim that the derived model not only allows for validation of the actual process against the design intent but also can be used for further process improvements. These improvements were mainly focused upon the resources available. The literature review did not reveal any additional attempts to apply Process Mining algorithms to industrial equipment logs until 2014 when Son et al. [8] presented their research into discovering process models for the product flow from the first step of manufacturing to final shipment. Due to a lag of detailed data, these models, however, cannot be used to enhance the performance of the individual machines involved. Yahya's research [9] was also focused on the path of the product through manufacturing. He noted that the granularity needs to be chosen and the process model customised to the analysis' goal. Yang et al. [10] propose the enhancement of such high-level production data with the help of unstructured data like emails. Although the technical approach is presented in detail, it remains unclear what added benefits such approach yields.

Farooqui et al. [11] realised that the implementation of additional code within industrial robot programs allow them to record more details relating to their work sequence. They are proposing to apply Process Mining algorithms to the resulting log to discover a matching process model which is helpful for maintenance work and also supports decision making. Brzychczy et al. [12] also see the benefits of utilising low-level machine data for their research. They indicate that one of the significant hurdles to overcome is the grouping of activities into cases. According to their work, this is best achieved through knowledge-based identification of the beginning and the end of a case. Nowaczyk et al. [13] are tapping into the 'wisdom of the crowd' by evaluating groups of peers. Deviates one of the observed systems from the behaviour of the remaining, similar systems, it can be concluded that maintenance is required.

3   Hypothesis

3.1   Log Pre-Processing

As shown in figure 1, a machine sequence can be split up into five distinct sections. A part being present in the machine is a signal common to all manufacturing equipment. Therefore the load event must be the event just before the part is present. Analog the unloading activity is observed while the part no longer is present in the machine. In cases where a machine manufactures multiple different parts, a setup, just prior to the load, might be required. If the same events always happen before a load event for a given part type, then it can be reasoned that those activities must be setup related. Finally, some machines require some additional motions so that the part can be unloaded. In figure 1 this is shown as pin 1 returning. This event has to be reset before the next part is loaded. Therefore a reset event is present if the same activity can be found after each unload event. Identifying the five sections described above allows for case identification within the equipment log.

Several hypotheses were devised that allow conclusions regarding the completeness of a case within an event log for automated production equipment. Within every cycle, there must be a load and an unload event. If a part is just passing through a station, it is even possible that those two activities are the only activities. Most non-robotic activities within the sequence have opposing motions. A typical example in figure 1 would be the closing and opening of clamp C01. If one of those two actions is missing, that could be an indicator for the logs incompleteness. Although robotic events do not have opposing motions, it is expected that a process follows the robots initiation and vice versa. Since log completeness is expected to be at a high level, it can be assumed that the most occurring trace class is complete. These knowledge-based rules can be used to annotate the log, and after manually tagging example logs, a machine learning model can be created which allows for the classification of the remaining cases.

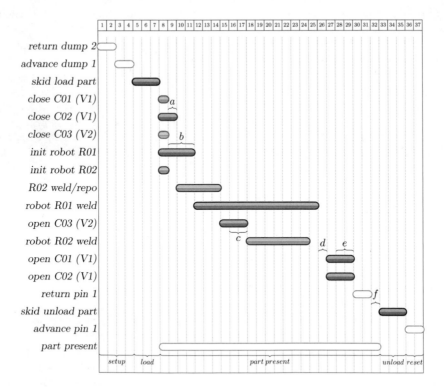

3.2   Automated Improvement Potential Detection

Excessive manual cycles: During production, the equipment typically is in automatic mode unless a problem occurs that requires manual intervention. The machines within an automotive body shop often are specified to provide an up-time of 80+%. If excessive manual cycles are recorded daily, it can be concluded that there is a systematical problem which needs to be addressed.

Identical units: Several pneumatic cylinders are often connected to a single solenoid. The grouped cylinders typically have the same bore and stroke and therefore should require the same time to advance and return. Setup can impact the synchronous movement of the units. This fact can be found in the event log. An example is shown in figure 1 where cylinders C01 and C02 are attached to the same valve, but their closing time is different. This improvement potential is marked in red and labelled with (a). With $\Delta\tau_{Se}$ being the duration of a station event, $\Delta\tau_{Se'}$ being the duration of an equivalent event triggered by the same solenoid and $\lambda\tau(Se_{ref})$ the mean duration of an identical reference event, setup problems are present if

$$\Delta\tau_{Se} \neq \Delta\tau_{Se'} \ \lor \ \Delta\tau_{Se} \neq \lambda\tau(Se_{ref}) \tag{1}$$

Opposing motions: If a motion in one direction takes longer than into the opposing direction, a setup problem is present as well. A nomenclature based algorithm can identify which activities are opposing motions. The open events for C01 and C02 in figure 1 take longer than their corresponding closing events. Therefore the potential improvement is labelled with (e). Let $\Delta\tau_{Se}$ be the duration of station event Se and $\Delta\tau_{\overline{Se}}$ the duration of the events opposing motion then the setup is correct if:

$$\Delta\tau_{Se} = \Delta\tau_{\overline{Se}} \tag{2}$$

Double triggers: If there are programming errors, an equipment motion may be started, interrupted and restarted again. Such behaviour causes increased cycle time and is responsible for excessive mechanical wear. In the log, this can be identified by an event which has a start timestamp but no complete timestamp followed shortly after by another event for the same activity that has both timestamps. Since events can happen twice within a case, the detection algorithm has to consider that the opposing motion did not happen in between. If the start timestamp of a station event Se is defined as $\tau_s(Se_n)$, the complete event as $\tau_c(Se_n)$ and the opposing motion of that event as $\overline{Se}_n$ then a double trigger is present if

$$\tau_s(Se_n) \neq \varnothing \ \land \ \tau_c(Se_n) = \varnothing \tag{3}$$

is followed by an identical event with

$$\tau_s(Se_{n+x}) \neq \varnothing \ \land \ \tau_c(Se_{n+x}) \neq \varnothing \tag{4}$$

as long as

$$Se_{n+1} \ldots Se_{n+(x-1)} \neq \overline{Se_n} \tag{5}$$

Bouncing motions: The term 'bouncing motion' was coined for a motion that reaches its end position but, due to the mechanical setup, bounces back so that it needs to be triggered once again to arrive at the stop position. In the event log, this can be identified by an event with start and complete timestamps followed shortly after by again the same event with start and complete timestamps without the opposing motion being recorded in between. Double triggers and bouncing motions manifest themselves in figure 1 similar to the opposing motion hypothesis mentioned previously (figure 1 (c)). However, the underlying data allow the discovery of the actual root cause. Based on above definitions a bouncing motion can be detected if

$$\tau_s(Se_n) \neq \varnothing \ \land \ \tau_c(Se_n) \neq \varnothing \tag{6}$$

is followed by an identical event with

$$\tau_s(Se_{n+x}) \neq \varnothing \ \land \ \tau_c(Se_{n+x}) \neq \varnothing \tag{7}$$

as long as

$$Se_{n+1} \ldots Se_{n+(x-1)} \neq \overline{Se_n} \tag{8}$$

Gaps: In the automotive body shop domain, there should be no period within a sequence, where there is no motion occurring. Considering that for this experiment, a variance of ~100ms within the timestamps was found, it can be concluded that any gap >200ms marks an area of possible improvement. Gaps can be caused either by programming errors or by external circumstances which are not recorded. A typical example of a gap is marked with the letter (d) within figure 1. Gaps can be detected by splitting up the timeline t of a case into bins. Then the number of events that fall within one such bin are counted and represented by ξ_t. Based on these definitions, a gap is present if

$$\sum_{t=x}^{x+200ms} \xi_t = 0 \tag{9}$$

with x being any value between the start timestamp $\tau_s(Se_i)$ of the incoming event Se_i and the start timestamp $\tau_s(Se_o)$ of the outgoing event Se_o of a case.

Station blocked: A particular case of the above described external circumstances, is the station being blocked. A blockage is caused by the next station not being ready to receive the completed part. In that case, the event data will show a gap before the unload event. A blocked condition has been highlighted within figure 1 with the letter (f). Let $\Delta\tau_p(Se_o)$ be the distance from the outgoing station even Se_o to the second last event then a blocked condition exists if

$$\Delta\tau_p(Se_o) > 0 \tag{10}$$

Special Event - Robot Initiation: The duration of the robot initiation process was found to be varying substantially. During this timeframe, the robot receives its program number and a start signal which triggers it to move to a pounce position. Typically this routine takes a maximum of two seconds what leads to the assumption that a robot initiation lasting more than two seconds is suspicious. Such a situation is shown in figure 1 with the letter (b). With $\Delta\tau_{Rinit}$ being the duration of a robot initiation event a reason for suspicion is present if

$$\Delta\tau_{Rinit} > 2sec. \tag{11}$$

4 Evaluation

4.1 Log Pre-Processing

To prepare the log for further processing the activities stemming from a double trigger or bouncing motion event, as described in chapter 3.2 were combined by merging the start timestamp of the first with the end timestamp of the second record. Next, the first activities, along with other, incomplete log items were removed. Python algorithms were developed to identify the five sections defining a case.

The five hypotheses introduced in section 3.1 created the basis for Python algorithms that can add engineering features to an industrial log. For 'load/unload present', 'station in bypass' and 'robot initiate & process present' a binary value of 0 or 1 was chosen. For the remaining features 'most occurring' and 'opposing motions present' a percentage, expressed as value between 0 and 1, was used. Manual tagging was performed for 500 random cases within the log available. Various machine learning algorithms, included in the Python scikit-learn package [14], were applied with cross-validation to the resulting matrix. Table 1 shows that a simple decision tree classifier, after tuning the hyperparameters, already achieves a 99% accuracy with +/-1% deviation. The associated confusion matrix, shown in figure 2, also exhibits no false positives.

4.2 Automated Improvement Potential Detection

For evaluation purposes, reasoning based algorithms for all of the above hypotheses (chapter 3.2) were implemented using Python and applied to event logs of

Table 1. Accuracy Of Classifier Models

	accuracy without hyper parameter tuning	accuracy with hyper parameter tuning
gradient boosting classifier	94% +/- 8%	99% +/- 1%
random forest classifier	93% +/- 8%	99% +/- 1%
decision tree	93% +/- 8%	99% +/- 1%
gaussian naive bayes	87% +/- 5%	87% +/- 5%
support vector classifier	91% +/- 8%	95% +/- 3%
logistic regression	91% +/- 12%	95% +/- 2%
k-nearest neighbors	90% +/- 7%	99% +/- 2%

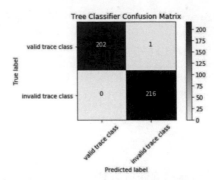

Fig. 2. The Confusion Matrix

four real-life framing re-spot stations. The issues automatically discovered for the four stations are summed up in the table 2.

Table 2. Evaluation Results For Four Body Shop Re-spot Stations

	station 1	station 2	station 3	station 4
number of cycles recorded	1157	1181	1185	1184
excessive manual cycles	< 0.5%	< 0.5%	< 0.5%	< 0.5%
time differences for identical units	0 sec.	0.2 sec.	0.2 sec.	0 sec.
time differences for opposing motions	1 sec.	2.3 sec.	1.2 sec.	1.9 sec.
double triggers events (# of times)	7 (88)	13 (1033)	9 (203)	11 (2045)
bouncing motions events (# of times)	0	2 (104)	0	0
gaps	0.8 sec.	6.4 sec.	10.7 sec.	0.4 sec.
station blocked	0.3 sec.	5.5 sec.	0.4 sec.	0.2 sec.
difference typical to fastest variant	0 sec.	0 sec.	0 sec.	10 sec.

5   Conclusion And Future Works

In a typical automotive body shop, one can almost always find a few stations which do not meet their expected throughput. This shortcoming might be due to long cycle times or increased maintenance activities. These stations are holding back the output of the body shop and therefore are termed 'bottlenecks'. Removing those few bottlenecks can increase the output of the body shop as a whole.

In this paper, hypotheses were formulated that allow for the automated generation of equipment logs and the subsequent discovery of hidden manufacturing potential. As proof, the prepositions were encoded and applied to real life equipment logs taken from bottleneck stations within an automotive body shop. The insight gained was automatically marked within Gantt charts. The results presented show that there is, for the evaluated stations, an improvement potential ranging from 2.1 seconds to 12.5 seconds. These findings lead to the conclusion that the analysis effort is worthwhile, even if it is assumed, that the current production process is well understood.

It is believed that more discoveries are possible. Therefore research will continue to focus on potential information gain based on more elaborate Process Mining techniques initially developed for business process analysis.

References

1. Koehler, W., Jing Y.: Automated, Nomenclature Based Data Point Selection for Industrial Event Log Generation. International Conference on Intelligent Data Engineering and Automated Learning, 31–40 (2018)
2. Lee, J.: E-manufacturing—fundamental, tools, and transformation. Robotics and Computer-Integrated Manufacturing. 501-507 (2003)
3. Lee, J., Bagheri, B., Jin, C.: Introduction to cyber manufacturing. Manufacturing Letters, 8. 11-15 (2016)
4. Jaber, A.A., Bicker, R.: The state of the art in research into the condition monitoring of industrial machinery. Int. J. of Current Engineering and Technology. 1986-2001 (2014)
5. Banerjee, T.P., Das, S.: Multi-sensor data fusion using support vector machine for motor fault detection. Information Sciences. 96-107 (2012)
6. Van der Aalst, W.M.: Process Mining: data science in action. Springer Berlin Heidelberg. (2016)
7. Hu, H., Li, Z. and Wang, A.: Mining of flexible manufacturing system using work event logs and petri nets. In International Conference on Advanced Data Mining and Applications 380-387 (2006)
8. Son, S., Yahya, B., Song, M., Choi, S., Hyeon, J., Lee, B., Jang, Y. and Sung, N.: Process Mining for manufacturing process analysis: a case study. In Proceeding of 2nd Asia Pacific Conference on Business Process Management (2014)
9. Yahya, B.N.: The development of manufacturing process analysis: lesson learned from Process Mining. Jurnal Teknik Industri 95-106 (2014)
10. Yang, H., Park, M., Cho, M., Song, M. and Kim, S.: A system architecture for manufacturing process analysis based on big data and Process Mining techniques. In 2014 IEEE International Conference on Big Data 1024-1029 (2014)
11. Farooqui, A., Bengtsson, K., Falkman, P. and Fabian, M.: From factory floor to process models: A data gathering approach to generate, transform, and visualize manufacturing processes. CIRP Journal of Manufacturing Science and Technology (2018).
12. Brzychczy, E. and Trzcionkowska, A.: Creation of an Event Log from a Low-Level Machinery Monitoring System for Process Mining Purposes. In International Conference on Intelligent Data Engineering and Automated Learning 54-03 (2018)
13. Nowaczyk, S., Sant'Anna, A., Calikus, E. and Fan, Y.: Monitoring Equipment Operation Through Model and Event Discovery. In International Conference on Intelligent Data Engineering and Automated Learning 41-53 (2018)
14. Pedregosa, F., Varoquaux, G., Gramfort, A., Michel, V., Thirion, B., Grisel, O., Blondel, M., Prettenhofer, P., Weiss, R., Dubourg, V. and Vanderplas, J.: Scikit-learn: Machine learning in Python. Journal of machine learning research 2825-2830 (2011)

Hardening Deep Neural Networks in Condition Monitoring Systems against Adversarial Example Attacks

Felix Specht and Jens Otto

[1] Fraunhofer IOSB-INA, Lemgo 32657, Germany
[2] https://www.cybersecurity-owl.de/
[3] {felix.specht,jens.otto}@iosb-ina.fraunhofer.de

Abstract. Condition monitoring systems based on deep neural networks are used for system failure detection in cyber-physical production systems. However, deep neural networks are vulnerable to attacks with adversarial examples. Adversarial examples are manipulated inputs, e.g. sensor signals, are able to mislead a deep neural network into misclassification. A consequence of such an attack may be the manipulation of the physical production process of a cyber-physical production system without being recognized by the condition monitoring system. This can result in a serious threat for production systems and employees. This work introduces an approach named CyberProtect to prevent misclassification caused by adversarial example attacks. The approach generates adversarial examples for retraining a deep neural network which results in a hardened variant of the deep neural network. The hardened deep neural network sustains a significant better classification rate (82% compared to 20%) while under attack with adversarial examples, as shown by empirical results.

1 Introduction

Cyber-physical production systems (CPPS) consist of hardware and software components controlling physical processes. They are in the focus of initiatives such as Germany's Industrie 4.0 or the US Industrial Internet Consortium. CPPS adapt efficiently to new products or product variants without extensive manual engineering effort [1–3].

Figure 1 a) shows an example of a CPPS where material is moved and processed between a storage module, a conveyer module, a heating module and a pick-and-place module.

Condition monitoring systems (CMS) can be utilized to detect system failures of CPPS (cf. figure 1 b), e.g. a broken heating module. Therefore, process data from modules are analyzed with machine learning algorithms such as deep neural networks (DNN) [4]. DNNs enable automatic generation of mathematical models representing the normal behavior of a CPPS. The normal behavior is learned using historical process data from production modules. As a representation of

© The Author(s) 2021
J. Beyerer et al. (Hrsg.), *Machine Learning for Cyber Physical Systems*, Technologien für die intelligente Automation 13,
https://doi.org/10.1007/978-3-662-62746-4_11

the learned normal behavior, the model is compared with the actual CPPS state to classify its condition as normal behavior or anomaly. DNNs have successfully been used to model physical manufacturing processes [5,6].

However, DNNs are vulnerable to adversarial example attacks [7]. Exploiting such attacks may result in a manipulation of the physical production process, which can cause enormous damage to facilities, production systems and employees [8].

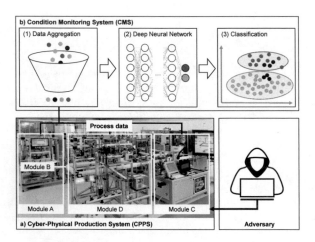

Fig. 1. Adversarial example attack against a condition monitoring system (b) results in misclassification of the observed cyber-physical production system (a).

An adversarial example (AE) is a specially manipulated input with the ability to mislead a DNN into misclassification [7]. It is generated from an undistorted original input by intentionally applying worst-case perturbations [9]. This results in an adversarial input being almost identical to the original one.

Fig. 1 shows an example of a condition monitored CPPS, where an adversary gained access to the production system. One objective of the adversary may be the manipulation of the production process without triggering an alert by the CMS (false-negatives). Another objective may be the triggering of false alerts (false-positives).

A successful attack may be achieved by the following steps: (i) process data from production modules is collected, (ii) collected process data is used to generate AEs, (iii) AEs are exploited to manipulate the physical process.

A false-positive classification triggers an anomaly alarm by the CMS, although the production process was actually correct. This may lead to unscheduled maintenance, which temporarily stops production. Likewise, the confidence in the CMS may be reduced. Furthermore, a false-negative classification may result in e.g. damaged products. A long-term operation in an insecure system state may even result in severe damage to the production system and pose a threat to employees [8].

Our contribution is the introduction of an approach to prevent misclassification caused by adversarial example attacks on deep neural network based condition monitoring systems, which detect system failures of cyber-physical production systems. Empirical results show that our approach results in a hardened deep neural network with a significant less misclassification rate despite being attacked.

2 Related work

Szegedy et al. introduce AEs as an anomalous property of a DNN [7]. A DNN can be formalized as a mathematical model $F(x, \theta) = Y$. The DNN decides whether a given input $x \in \mathbb{R}^n$ belongs to a learned class $Y \in \mathbb{R}^m$, using a set of internal parameters θ.

An AE x' is defined as an incorrectly classified input, which deviates minimally from the correctly classified original x. As shown in definition 1, an AE is generated by applying perturbations Δ_x to the original input x, where Δ_x is kept as small as possible.

$$\text{find } x' = x + \Delta_x \tag{1}$$
$$\text{s.t. } F(x', \theta) = Y', \ Y' \neq Y$$

A quality criterion for AEs is inconspicuousness and a minimal deviation from its original. E.g. in image classification an objective of AEs is the perturbation to the original, imperceptible to the human eye. For quantification of this property the three distance metrics L_0, L_2 and L_∞ are commonly used in literature [10]. The L_0 metric corresponds to the number of input signals that have been altered (e.g. pixels). The L_2 metric measures the standard Euclidean (root-mean-square) distance between x and x'. At last, the L_∞ metric measures the maximum change for any of the input signals.

Several approaches for generating AEs have been proposed, such as box-constrained limited-memory Broyden-Fletcher-Goldfarb-Shanno (L-BFGS), Fast Gradient Sign Method (FGSM) and Projected Gradient Descent (PGD) [7,9,11]. By leveraging AE generating approaches, both Black-Box attacks and White-Box attacks against machine learning classifiers can be performed [12].

Several defensive strategies have been proposed to harden a DNN model against AE attacks, such as Adversarial Training, Defensive Distillation, feature squeezing and PGD based Adversarial Training [7,11,13,14].

This work transfers AE generation and prevention into the field of industrial automation, in contrast to presented approaches considering image processing mainly. Our approach adapts the FGSM approach [9] for AE generation and the adversarial training [7] for preventing misclassification. Both, FGSM and adversarial training are suitable for the CPPS requirement of rapid adaptability.

3 Solution

The objective of the CyberProtect approach is prevention of misclassification caused by AE attacks and is achieved by the following steps: (i) A DNN is exclusively trained on process data $\mathcal{P}$ in an initial training phase. (ii) The training phase is extended by an a additional retraining phase, where $\mathcal{P}$ is used to generate an AE $\mathcal{P}'$ as described in the previous section. (iii) Both, the original $\mathcal{P}$ and the manipulated $\mathcal{P}'$ serve as input to the DNN. Generating the required AEs can be formalized as follows.

Process data of a CPPS is defined as $\mathcal{P} = (p_0, ..., p_m)$, where $p_i \in [0, 1]$ for $i \in \{0, ..., m\}$ is a sensor or actor value. $\mathcal{P}$ is input to the CMS utilizing a DNN. As described in section 2, a DNN is a mathematical model $F(\mathcal{P}, \theta) = Y$, where $Y \in \{0, 1\}$ is the predicted class corresponding to normal or anomaly behavior and θ is a set of parameters. The adversarial objective is reached by solving the following search problem:

$$\text{find} \quad \mathcal{P}' = \mathcal{P} + \Delta \tag{2a}$$

$$\text{s.t.} \quad F(\mathcal{P}, \theta) = Y \tag{2b}$$

$$F(\mathcal{P} + \Delta, \theta) = Y' \neq Y \tag{2c}$$

$$\sqrt{\sum_{i=0}^{m} |\delta_i|^2} \leq \epsilon, \ \epsilon > 0 \tag{2d}$$

A perturbation $\Delta = (\delta_0, ..., \delta_m)$, where $\delta_i \in [-1, 1]$ for $i \in \{0, ..., m\}$, is added to an original $\mathcal{P}$ to generate an adversarial example $\mathcal{P}' = (p_0 + \delta_0, ..., p_m + \delta_m)$. Due to constraint 2c, $\mathcal{P}'$ is not predicted as the original class 2b. The constraint 2d increases the inconspicuousness of $\mathcal{P}'$ by limiting the euclidean distance between $\mathcal{P}'$ and $\mathcal{P}$ to an upper bound.

By exploiting adversarial examples, the adversary can manipulate the production process unrecognized within the specified constraints. This leads to false-positive or false-negative results in classification by the condition monitoring system.

3.1 Generation of Adversarial Examples Algorithm

This approach generates AEs by using the Fast Gradient Sign Method (FGSM) [9]. Due to its low computational costs, FGSM is suitable for the CPPS requirement of rapid adaptability.

Generation of AEs is formally described by algorithm 1. The algorithm requires inputs $\mathcal{P}, F(\mathcal{P}, \theta), Y, \epsilon, s$, where $\mathcal{P}$ describes process data, $F(\mathcal{P}, \theta)$ describes the trained DNN, Y is the original class label, ϵ is a threshold parameter and s is a precision parameter.

The algorithm performs the following steps: (1) A variable η is increased by the precision parameter s specifying the growth of η between iterations. (2) FGSM is used to generate a candidate for an AE $\mathcal{P}'$. FGSM requires $F(\mathcal{P}, \theta)$

Algorithm 1: GenAE: Generation of Adversarial Example $\mathcal{P}'$ from original Process Data $\mathcal{P}$

Input: Process data $\mathcal{P}$, DNN $F(\mathcal{P}, \theta)$, class label Y, threshold parameter ϵ, precision parameter s

Output: Adversarial example candidate $\mathcal{P}'$

1   $\eta \leftarrow 0$
   `// while misclassification` Y' `not found and euclidean distance`
      `threshold` ϵ `not reached`
2   **while** $Y' = Y$ ***and*** $\|\mathcal{P}' - \mathcal{P}\| \leq \epsilon$ **do**
     `// increase perturbation`
3      $\eta \leftarrow \eta + s$
     `// use Fast Gradient Sign Method to generate an AE candidate`
4      $\mathcal{P}' \leftarrow FGSM(F(\mathcal{P}, \theta), \eta)$
     `// get class label`
5      $Y' \leftarrow F(\mathcal{P}', \theta)$
6   **return** $\mathcal{P}'$

---

and the variable η and extracts the gradients from a previous trained DNN $F(\mathcal{P}, \theta)$. (3) The class Y' is predicted by the trained DNN, where the generated candidate $\mathcal{P}'$ serves as input. These three steps are repeated until either the euclidean distance between the candidate $\mathcal{P}'$ and the original process data $\mathcal{P}$ exceeds the threshold parameter ϵ, or the new predicted class label Y' differs from the original class label Y. In the case of differing classes, a valid AE is found. The algorithm returns the last computed AE candidate $\mathcal{P}'$.

3.2   CyberProtect Algorithm

The CyberProtect algorithm 2 enables prevention of misclassification caused by AE attacks. The algorithm requires a $\mathcal{P}^n, Y^n, \theta, \epsilon, s$ as input. Input $\mathcal{P}^n$ describes n-dimensional historical process data $\mathcal{P}$, Y^n describes the corresponding n-dimensional class labels, θ describes a DNN configuration, ϵ and s are configuration parameters to use GenAE algorithm described above.

The algorithm executes the following steps: (1) A DNN $F(\mathcal{P}, \theta)$ is initialized with the configuration parameters θ (cf. line 1 function *initialize*) for DNN architecture, activation and cost function. (2) the DNN is trained with each entry $\mathcal{P}_i$ and Y_i (cf. line 2-3 function *train*) of the historical process data $\mathcal{P}^n$ and the corresponding class labels Y^n. (3) An empty set $\mathcal{P}'^n$ is defined (cf. line 4) after the first training phase. (4) The algorithm GenAE 1 is used to generate and store AEs to $\mathcal{P}'^n$ (cf. line 5-7) for each process data entry $\mathcal{P}_i$ resulting in an AE $\mathcal{P}'_i$. (5) A new DNN $\hat{F}(\mathcal{P}, \theta)$ is initialized with the configuration parameters θ (cf. line 8). (6) $\hat{F}(\mathcal{P}, \theta)$ is trained with each entry $\mathcal{P}_i$ and $\mathcal{P}'_i$ of both, the original process data $\mathcal{P}^n$ and the generated AEs $\mathcal{P}'^n$ (cf. line 9-11) using the same class label Y_i for $\mathcal{P}_i$ and $\mathcal{P}'_i$. The algorithm returns the new trained DNN $\hat{F}(\mathcal{P}, \theta)$ hardened against AE attacks.

Algorithm 2: CyberProtect

Input: Process data set $\mathcal{P}^n$, set of class labels Y^n, configuration parameter θ,
        threshold parameter ϵ, precision parameter s
Output: Hardened Deep Neural Network $\hat{F}(\mathcal{P}, \theta)$

1   $F(\mathcal{P}, \theta) \leftarrow initialize(\theta)$
   // train DNN with process data set $\mathcal{P}^n$
2   **for** $i = 1$ *to* n **do**
3       $F(\mathcal{P}, \theta) \leftarrow train(\mathcal{P}_i, Y_i)$
4   $\mathcal{P'}^n \leftarrow \emptyset$
   // calculate AE set from process data
5   **for** $i = 1$ *to* n **do**
       // cf. section 3.1
6       $\mathcal{P}'_i \leftarrow GenAE(\mathcal{P}_i, F(\mathcal{P}, \theta), Y_i, \epsilon, s)$
7       $\mathcal{P'}^n \cup \{\mathcal{P}'_i\}$
8   $\hat{F}(\mathcal{P}, \theta) \leftarrow initialize(\theta)$
   // train DNN with process data set $\mathcal{P}^n$ and AE set $\mathcal{P'}^n$
9   **for** $i = 1$ *to* n **do**
10      $\hat{F}(\mathcal{P}, \theta) \leftarrow train(\mathcal{P}_i, Y_i)$
11      $\hat{F}(\mathcal{P}, \theta) \leftarrow train(\mathcal{P}'_i, Y_i)$
12 **return** $\hat{F}(\mathcal{P}, \theta)$

---

4   Results

Emperical results were obtained with a reference CMS monitoring data from the Secom dataset [15]. The Secom dataset was recorded from a semi-conductor manufacturing process and consists of process data with 590 attributes collected from sensor signals and variables during 1567 manufacturing cycles.

The reference CMS is implemented based on a DNN by using the python framework tensorflow [16]. The applied DNN architecture consists of 590 input neurons, 4 hidden layer with 590, 1180, 2360, 590 neurons each and one output neuron representing the conditions normal or anomaly. As activation function rectified linear unit (ReLU) is applied. Training the DNN was performed in a supervised manner for 1000 epochs using the Adam optimizer [17] with parameters $\beta_1 = 0.9$, $\beta_2 = 0.999$, $\epsilon = 10^{-8}$ and a learning rate of 0.01.

Our CyberProtect implementation extends the reference CMS and is based on the Python library Cleverhans [18], an extension to the Tensorflow framework [16].

The reference CMS was extended with the CyberProtect algorithm to obtain results shown in Fig. 2.

The left column of Fig. 2 shows classification results from the CMS reference excluding CyberProtect. The mean classification rate is equal to 82% with a standard deviation of 7%, the best result is 95% and the worst result is 60%. The middle column shows classification results of the same CMS reference under AE attacks based on FGSM generation. The classification rate is reduced to a

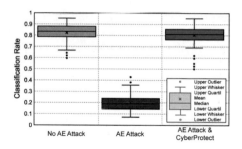

Fig. 2. Classification rate of a common CMS without AE attacks (left), with AE attacks (middle) and a CMS using CyberProtect with AE attacks (right).

mean of 20% with a standard deviation of 7%, best result of 43% and worst result of 7% . In the right column, results are shown of the extended reference CMS utilizing the CyberProtect approach while being attacked with AEs. Here, CyberProtect significantly increases the classification rate 80% with a standard deviation of 9%, best result of 95% and worst result of 50%.

CyberProtect enables a DNN to nearly regain the classification rate despite AE attacks, as demonstrated in carefully designed experiments.

5   Conclusion

This paper presents the CyberProtect approach to prevent misclassification caused by adversarial example attacks on deep neural network based condition monitoring systems in the domain of cyber-physical production. Adversarial example attacks can result in a serious threat to production systems and employees, due to their ability to manipulation the monitored production process unrecognized.

This work formally defines generation of adversarial examples as a constrained search problem and uses adversarial examples to retrain a deep neural network. Empirical results prove that a deep neural network hardened by CyberProtect show a significant less misclassification rate despite being attacked.

In future work, prevention of misclassification caused by adversarial example attacks will be explored for discrete manufacturing, in which time-dependent machine learning approaches are utilized.

References

1. J. Otto, B. Vogel-Heuser, and O. Niggemann.   Online parameter estimation for cyber-physical production systems based on mixed integer nonlinear programming, process mining and black-box optimization techniques.   *at-Automatisierungstechnik*, 66(4):331–343, 2018.
2. G. Reinhart, S. Krug, S. Huttner, Z. Mari, F. Riedelbauch, and M. Schlogel. Automatic configuration (plug & produce) of industrial ethernet networks. In *Proc.*

9th IEEE/IAS International Conference on Industry Applications (INDUSCON), pages 1–6, Sao Paulo, Brazil, nov 2010.

3. J. Otto, B. Vogel-Heuser, and O. Niggemann. Automatic parameter estimation for reusable software components of modular and reconfigurable cyber-physical production systems in the domain of discrete manufacturing. *IEEE Transactions on Industrial Informatics*, 14(1):275–282, 2018.

4. L. Monostori, B. Kádár, T. Bauernhansl, S. Kondoh, S. Kumara, G. Reinhart, O. Sauer, G. Schuh, W. Sihn, and K. Ueda. Cyber-physical systems in manufacturing. *International Academy for Production Engineering Annals*, 65(2):621–641, 2016.

5. D. Hossain, G. Capi, M. Jindai, and S. Kaneko. Pick-place of dynamic objects by robot manipulator based on deep learning and easy user interface teaching systems. *Industrial Robot: the international journal of robotics research and application*, 44(1):11–20, 2017.

6. S. Jeschke, C. Brecher, T. Meisen, D. Özdemir, and T. Eschert. *Industrial internet of things and cyber manufacturing systems*, pages 3–19. Springer International Publishing, 2017.

7. C. Szegedy, W. Zaremba, I. Sutskever, J. Bruna, D. Erhan, I. Goodfellow, and R. Fergus. Intriguing properties of neural networks. In *Proc. of the 2nd International Conference on Learning Representations (ICLR)*, Banff, Canada, apr 2014.

8. K. Stouffer, J. Falco, and K. Scarfone. Guide to industrial control systems (ics) security. *NIST special publication*, 800(82):16, 2011.

9. I. Goodfellow, J. Shlens, and C. Szegedy. Explaining and harnessing adversarial examples. In *Proc. of the 3rd International Conference on Learning Representations (ICLR)*, San Diego, USA, may 2015.

10. N. Carlini and D. Wagner. Towards evaluating the robustness of neural networks. In *Proc. of the 38th IEEE Symposium on Security and Privacy (SP)*, pages 39–57, San Jose, USA, may 2017.

11. Aleksander Madry, Aleksandar Makelov, Ludwig Schmidt, Dimitris Tsipras, and Adrian Vladu. Towards deep learning models resistant to adversarial attacks. *arXiv preprint arXiv:1706.06083*, 2017.

12. N. Papernot, P. McDaniel, and I. Goodfellow. Transferability in machine learning: from phenomena to black-box attacks using adversarial samples. *Computing Research Repository (CoRR)*, abs/1605.07277, 2016.

13. N. Papernot and P. McDaniel. On the effectiveness of defensive distillation. *Computing Research Repository (CoRR)*, abs/1607.05113, 2016.

14. W. Xu, D. Evans, and Y. Qi. Feature squeezing: Detecting adversarial examples in deep neural networks. *Computing Research Repository (CoRR)*, abs/1704.01155, 2017.

15. M. McCann and A. Johnston. Uci ml repository secom dataset, 2008. [Online; accessed 2018-02-05].

16. M. Abadi et al. Tensorflow: A system for large-scale machine learning. In *Proc. of the 12th USENIX Symposium on Operating Systems Design and Implementation (OSDI)*, volume 16, pages 265–283, Savannah, USA, nov 2016.

17. DP. Kingma and J. Ba. Adam: A method for stochastic optimization. *Computing Research Repository (CoRR)*, abs/1412.6980, 2014.

18. I. Goodfellow, N. Papernot, and P. McDaniel. cleverhans v2.0.0.: an adversarial machine learning library. *Computing Research Repository (CoRR)*, abs/1610.00768, 2016.

First Approaches to Automatically Diagnose and Reconfigure Hybrid Cyber-Physical Systems

Alexander Diedrich[1][0000−0002−8674−6895], Kaja Balzereit[1][0000−0001−9203−5902], and Oliver Niggemann[2]

[1] Fraunhofer IOSB-INA, Fraunhofer Center for Machine Learning, Lemgo, Germany
`surname.name@iosb-ina.fraunhofer.de`
[2] Helmut-Schmidt University, Hamburg, Germany
`oliver.niggemann@hsu-hh.de`

Abstract. Maintaining modern production machinery requires a significant amount of time and money. Still, plants suffer from expensive production stops and downtime due to faults within individual components. Often, plants are too complex and generate too much data to make manual analysis and diagnosis feasible. Instead, faults often occur unnoticed, resulting in a production stop. It is then the task of highly-skilled engineers to recognise and analyse symptoms and devise a diagnosis. Modern algorithms are more effective and help to detect and isolate faults faster and more precise, thus leading to increased plant availability and lower operating costs.

In this paper we attempt to solve some of the described challenges. We describe a concept for an automated framework for hybrid cyber-physical production systems performing two distinct tasks: 1) fault diagnosis and 2) reconfiguration. For diagnosis, the inputs are connection and behaviour models of the components contained within the system and a model describing their causal dependencies. From this information the framework is able to automatically derive a diagnosis provided a set of known symptoms. Taking the output of the diagnosis as a foundation, the reconfiguration part generates a new configuration, which, if applicable, automatically recovers the plant from its faulty state and resumes production. The described concept is based on predicate logic, specifically Satisfiability-Modulo-Theory. The input models are transformed into logical predicates. These predicates are the input to an implementation of Reiter's diagnosis algorithm, which identifies the minimum-cardinality diagnosis. Taking this diagnosis, a reconfiguration algorithm determines a possible, alternative control, if existing. Therefore the current system structure described by the connection and component models is analysed and alternative production plans are searched. If such an alternative plan exists, it is transmitted to the control of the system. Otherwise, an error that the system is not reconfigurable is returned.

Keywords: Model-based Diagnosis · Fault detection and isolation · Reconfiguration.

J. Beyerer et al. (Hrsg.), *Machine Learning for Cyber Physical Systems*, Technologien für die intelligente Automation 13,
https://doi.org/10.1007/978-3-662-62746-4_12

1   Introduction

Modern production machinery shall act as autonomously as possible [1]. Autonomous machines are characterized by the capability of making and implementing their own decisions regarding resource use and utilization of components. This also includes the capabilities of observing their own behaviour (self-monitoring), diagnosing their faults (self-diagnosis), and restoring valid system behaviour in case of faults (self-healing) [5].

However, there is still no holistic diagnosis and reconfiguration method which can successfully deal with heterogeneous production plant data and the resulting complex models. Most available diagnosis and reconfiguration methods instead tackle sub-problems, such as system modelling, diagnosis logical circuits, or reconfiguration in narrow and controlled domains. Therefore, developing and realising a robust method is still a research challenge.

Performing consistency-based diagnosis is the only known method to realistically find faults in complex systems. Other approaches are heuristic or require every possible system behaviour to be modelled exhaustively. Heuristic models model behaviour against the flow of causality. Based on an input (the effect) they calculate the likely cause of the fault. Models are created in a data-driven way and require a sufficient amount of training examples for each possible output. This amount of training data is often unavailable. Approaches that model every possible system behaviour exhaustively are often limited by real-world constraints such as the unavailability of accurate enough models.

Consistency-based diagnosis brings the advantage that only the normal behaviour of a system needs to be modelled. Thus, no adversarial examples need to be produced (as would be required for heuristic models) and engineers do not need to think of and simulate all possibilities how components can fail. This decreases modelling effort and avoids errors within the model. Additionally it has an advantage over heuristic models as it reasons with the flow of causality using a combination of deduction and abduction. Deduction propagates values from the system input to the system output and shows normal system behaviour. Through abduction deviations from this normal behaviour can be traced back to sets of components which are likely to have caused the fault. Abduction is more similar to the way humans diagnose systems. They analyse the faulty output and then look through the system from the output back to the input until they have identified components whose faulty behaviour might have caused the faulty output.

Figure 1 shows the general concept of the diagnosis and reconfiguration framework. The physical production plant generates process data from its sensors. This process data is discretised in the form of symptoms. A symptom of a signal shows the direction of deviation from normal behaviour (*high*, *low*, or *normal*). Additionally, experts need to provide two kinds of models: a connection model and individual component models. The component models numerically describe the behaviour of each component. For example, a water tank would be modelled by a difference equation. These component models need to be available for each piece of diagnosable equipment within the plant. Further, a system description

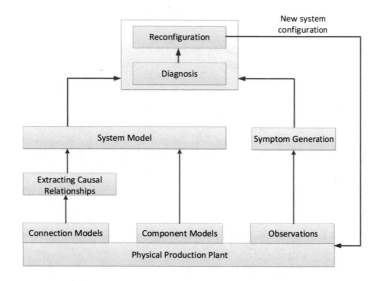

Fig. 1. Overview of the automated diagnosis and reconfiguration framework

in the form of a connection model needs to be provided. The connection model is a directed graph showing the causal connections between the employed components. A diagnosis algorithm [15], given the component and connection models and a set of symptoms, computes the smallest amount of possible faulty components that explain the symptoms. To obtain these models it is conceivable to create a digital twin during inception and construction of the plant. The data from the digital twin, such as simulation data, can be extracted and transformed into predicate logic models.

The set of possible faulty components is the input to a reconfiguration algorithm. For reconfiguration, the algorithm takes the structural and component models into consideration [2]. The algorithm does not only search for a new parametrization of the system but also looks for alternative paths that can be used to bypass the faulty components. From these it generates alternative control sequences, which reconfigure plant parameters or use redundant components to keep the plant operating.

We formulate the connection and component models through logical approaches to perform consistency-based diagnosis. Given proper models, we assume the set of symptoms as given. The symptoms can be generated through the use of well-known machine-learning methods such as principal component analysis and artificial neural networks. Diagnosis is realised through an implementation of Reiter's algorithm [15]. The reconfiguration method is based on a combination of causal reasoning and numerical parametrization approaches.

2   State of the Art

Struss [18] published a paper on the fundamentals of MBD of dynamic systems. In this he described how hybrid systems can be modelled without resorting to a complete simulation of the system under investigation. He proposed to capture the temporal and dynamic behaviour of a hybrid system in a set of modes which model the system. Each mode has distinct state and temporal constraints in addition to so called Continuity, Integration, and Derivatives (CID) constraints that affect all modes.

Daigle et al. [4] have adapted a discrete event approach to diagnose continuous systems. They state that each fault that occurs in a continuous system has a unique fault signature. A fault signature denotes a qualitative effect that a fault occurs in an observation. Under the assumption that all fault signatures and measurement orderings are known, they employ a diagnoser that traces the states through a temporal causal graph based on measurements.

Roychoudhury et al. [16] have shown how to use hybrid bond graphs (HBG) to diagnose hybrid systems. HBGs abstractly model the system by describing causal, continuous relationships between components. Daigle et al. [4] have employed the developed HBGs to diagnose a spacecraft power distribution system. Prakash et al. [13] have used an extended framework with HBGs to make improvements in diagnosing two-tank systems.

Grastien [8] used SMT for the diagnosis of hybrid systems. He discretizes values in a hybrid system into a set of distinct states. Each observation $< \tau, A >$ is understood as a behaviour A at time τ, where A is a partial assignment of the variables in a state. Each variable is augmented with an indicator stating at which time-step the variable expression is valid.

Fränzle et al. [7] have augmented SMT with probabilistic approaches in order to analyse stochastic hybrid systems. By using bounded-model checking together with probabilistic hybrid automata, piecewise deterministic Markov processes, and stochastic differential equations they are able to create a fault analysis system without the need to formulate intermediate finite-state abstractions as the methods mentioned above do.

In another work, Khorasgani [10] describe a hybrid system model through hybrid minimal structurally overdetermined sets (HMSOs). These are sets of differential equations and (in-) equations which model the behaviour of a hybrid system.

Crow et al. [2] extended Reiter's diagnosis algorithm so it is also capable of determining the components that need to be reconfigured. The components that need reconfiguration are determined in an analogous way as diagnosis is done.

Kobi et al. [11] presented an approach, how to identify and modify the process input. In case of parameter variations, the control input to the system is adapted.

Hwang et al. [9] published a survey on existing fault detection and reconfiguration methods: Most of the existing approaches rely on a quantitative analysis of the system data. Therefore, the numerical values of the system are analyzed. Structural information like a system topology are either not considered or implemented statically into the method.

Fleischanderl et al. [6] and Sabin et al. [17] presented configuration approaches based on constraint satisfaction. The configuration problem, which is to find an assembly of production tasks given product and production requirements, is mapped onto a constraint satisfaction problem, which task is to find a valid variable assignment subject to some given constraints.

In contrast to Struss [18] and Provan [14] we do not use automatons and mode estimation to partition the system into different states. Instead, we only sample the system at some suitable interval and use the obtained information directly to model the states in the state-space representation. Unlike in space-craft, which where analyzed by Daigle [3], fault signatures and measurement orderings are unknown in industrial systems. This requires us to pursue a more uninformed approach. Our approach is an alternative to hybrid bond graphs used by Roy-choudhury [16], while they are at the same time an extension to the work of Grastien [8] and Khorasgani [10]. In comparison to Grastien we do not singly use satisfiability modulo theory, but instead capture system behaviour in a state-space representation. We expect this to reduce the required computational effort. We also make use of (in-) equations and differential equations as were used by Khorasgani and Biswas, but augment these with the diagnostic reasoning of traditional model-based diagnosis. Compared to Fränzle, we do not make use of stochastic SMT at this point to keep the system more explainable for users.

3   The multiple-tank model

For this work we will use the four tank system depicted in Figure 2 as a running example. The system consists of four water tanks t, seven electric valves v

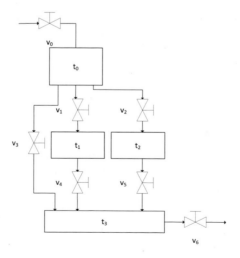

Fig. 2. The demonstration use case showing a four-tanks model

with integrated flow sensors, an unlimited water source and an unlimited water sink (not shown). Valve v_0 controls water from the unlimited water source, for example the public water mains, into tank t_0. From there, three pipes with an equal diameter divide the water flow. Finally, valve v_6 drains tank t_3 into the unlimited water sink, for example a river or a processing facility.

Each tank has two binary sensors which indicate overflow and underflow, respectively. There are no provisions to directly measure the water level. Each valve has a switch which indicates whether or not the valve is open. In addition, each valve has an associated flow sensor.

4   Diagnosing Hybrid Systems

Automatically diagnosing real hybrid systems is a hard task. So far, the only known diagnosis method which can deal with this kind of complexity is consistency-based. The method works by reasoning against the flow of causality, meaning that it uses abduction to determine likely fault causes by evaluating observations. The drawback of this kind of diagnosis is its reliance on accurate models. This diagnosis method requires the availability of three types of inputs. The component models (CM), a connection model (CON), and observations (OBS). In most plants component models are not available and must be obtained through expert knowledge. In the four-tank model and within the process industry in general, these components models are often differential equations or piece-wise functions. The water level in tanks and other tank-like components can be modelled through equations such as

$$\Delta h = \frac{1}{A}(Q_i + C_d a \sqrt{2gh_0}) \qquad (1)$$

and discrete switching signals (f.e. from valves) can be modelled with piece-wise functions such as

$$o(h_i, \tau_i^o) = \begin{cases} 0 & \text{if } h_i \leq \tau_i^o \\ 1 & \text{else} \end{cases}. \qquad (2)$$

The challenge is to obtain these models automatically. Often this can only be done with data-driven models such as max-margin approaches, artificial neural networks, or statistical methods.

The second kind of model is the connection model. CON is a directed graph. The nodes are the individual components whose input and outputs are governed by the component models. The edges of the directed graph show which component input values are related to which component output values. The graph can be obtained by fusing connection information automatically extracted from piping and instrumentation diagrams (P&I) with the component models.

A consistency-based diagnosis algorithm uses the models to calculate the normal system state and merges this information with the observations to perform fault identification. To merge these disparate models they need to be transformed into a suitable logical framework. For this, we have used Satisfiability Modulo

Linear Arithmetic (SMT-$\mathcal{LRA}$). The logic framework allows the formulation of equations, in-equations, and logical predicates as rules and store them in a knowledge base. Within each rule a component is identified through it's health state denoted as $ok(c_i)$. Given CON, an algorithm can determine what effects the failure of a component can have on its surrounding components. Here we focus on describing only normal behaviour. One logical rule for valve 2 in the running example is

$$ok(v_2) \wedge ok(t_0) \wedge ok(t_1) \rightarrow ok(f_2) \tag{3}$$

where v_2 is valve 2, t_0, t_1 are the adjacent tanks, and f_2 is a flow sensor for v_2. All rules have ok-assumptions for components on the left-hand side and observations on the right-hand side. Reading the rule from left to right uses deduction and tells the algorithm the normal state of the system: "if all components are ok, the flow sensor will show ok readings". For diagnosis the algorithm uses deduction: "Given that the flow f_2 is not ok, the components on the left-hand side are likely candidates". When the rules are created it must be ensured that rules have overlapping sets of components. Otherwise single components cannot be discriminated.

The transformation from the component model into ok-assumptions can be done through standard machine learning algorithms or be integrated into the logical framework itself. For example, the flow through valve 2 can either be calculated using the equations governing the inflow and outflow of tank t_0, or a simple machine learning algorithm can be trained which outputs ok/nok.

We employ Reiter's diagnosis algorithm [15] to evaluate the generated knowledge base and discriminate faults to obtain a diagnosis that contains the smallest amount of components (minimal cardinality diagnosis).

5 Reconfiguration after faults occurred

After a fault in the system is identified, a reconfiguration method is used to restore valid system behaviour - if possible. So a reconfiguration method works on a sophisticated level it needs to satisfy different requirements: First, the reconfiguration must be done in a short time with a minimal manual effort. Plant downtimes and component failures need to be minimized so that the costs of these errors are reduced. Additionally, the control software needs to adapt to different product specifactions and production modes. Therefore, it must not be static but needs to be able to adapt dynamically. It also needs to handle the complexity of the production plant and therefore consider the system parametrization as well as the system topology simultaneously. A lot of research has been done on the dynamic optimization of the numeric parameters of a system [12, 20, 19]. However, most of these methods only work for a static system configuration and cannot adapt to varying demands. The here presented approach differs from the state of the art since it considers both, the system parametrization and the system topology simultaneously. Thus, complex systems and varying production demands can be handled by the reconfiguration method. A reconfiguration method also needs to be able to separate between valid and invalid plant behaviour. In general, this information relies on expert knowledge: To separate faulty from non-faulty system behaviour, models representing non-faulty system behaviour are trained based on a set of non-faulty system data. This set has to be determined by an expert. Non-faulty system behaviour may also be invalid, if it does not lead to the required system goal. Thus, an expert has to define the current system goal to make sure that the current configuration leads to this goal.

A reconfiguration method takes a system description consisting of connection model and component models, the current system's state and a definition of the system goal as input. The connection model and the component models are the same as those used for diagnosis. The definition of the system goal is used to determine which system behaviour leads to the correct system goal. For the tank model, a possible system goal is to fill tank t_3. Every system behaviour leading to t_3 not being filled is invalid. The goal of reconfiguration is to restore valid system behaviour after a fault occurred.

For a reconfiguration the system's properties like its modules and their interconnections are modelled as well as the system goal. The current system state is checked. If it is valid, no action is necessary; the current system behaviour leads to the required production goal. If the current system state is invalid, the necessary system actions to restore valid system behaviour are determined.

Assuming that the pipe connecting t_0 and t_3 is broken and has been identified, based on a diagnosis, as faulty component. The reconfiguration method now returns the control instruction, that the pipe no longer should be used and proposes the connections (v_1, t_1, v_4) and (v_2, t_2, v_5) as alternatives so that tank t_3 is filled.

6   Conclusion and future work

The presented concept considers the automated diagnosis and reconfiguration of hybrid cyber-physical systems. Based on the current system data, a connection model representing the current structure of the system and component models, which model the behaviour of every component, a diagnosis is executed. Thus a set of possible root causes is determined.

After that a reconfiguration method is started: The task of reconfiguration is to restore valid system behaviour after a fault occurred. Given a system goal, the necessary actions for the recovery of the system are determined. Alternative production paths and parametrizations are identified so that the specified system goal still can be reached.

Our future work will be focussed on the automatic extraction of expert knowledge from P&I diagrams and learning components models from data. Currently, the connection model is extracted manually from the system structure. However, this is time consuming and requires a lot of manual effort. To reduce this, the needed information shall be extracted automatically. Creating component models automatically requires even more research effort. The behaviour of physical components is often governed by differential equations sometimes including non-linearities. For accurate component models this behaviour needs to be learned and accurately predicted by data-driven models. What models to use and how to train them remains a research challenge.

7 Acknowledgement

This work was developed within the Fraunhofer Cluster of Excellence "Cognitive Internet Technologies".

References

1. Die Bundesregierung. Strategie Künstliche Intelligenz der Bundesregierung. 2018.
2. Judith Crow and John M Rushby. Model-based reconfiguration: Toward an integration with diagnosis. In *AAAI*, pages 836–841, 1991.
3. Matthew Daigle, Xenofon Koutsoukos, and Gautam Biswas. A discrete event approach to diagnosis of continuous systems. In *Proceedings of the 18th International Workshop on Principles of Diagnosis*, pages 259–266, 2007.
4. Matthew J Daigle, Indranil Roychoudhury, Gautam Biswas, Xenofon D Koutsoukos, Ann Patterson-Hine, and Scott Poll. A comprehensive diagnosis methodology for complex hybrid systems: A case study on spacecraft power distribution systems. *IEEE Transactions on Systems, Man, and Cybernetics-Part A: Systems and Humans*, 40(5):917–931, 2010.
5. European Factories of the Future Research Association. Multi-Annual Roadmap for the Contractual PPP under HORIZONS 2020, 2013.
6. Gerhard Fleischanderl, Gerhard E Friedrich, Alois Haselböck, Herwig Schreiner, and Markus Stumptner. Configuring large systems using generative constraint satisfaction. *IEEE Intelligent Systems and their applications*, 13(4):59–68, 1998.
7. Martin Fränzle, Holger Hermanns, and Tino Teige. Stochastic satisfiability modulo theory: A novel technique for the analysis of probabilistic hybrid systems. In *International Workshop on Hybrid Systems: Computation and Control*, pages 172–186. Springer, 2008.
8. Alban Grastien. Diagnosis of hybrid systems by consistency testing. In *24th International Workshop on Principles of Diagnosis (DX-13)*, pages 9–14. Citeseer, 2013.
9. Inseok Hwang, Sungwan Kim, Youdan Kim, and Chze Eng Seah. A survey of fault detection, isolation, and reconfiguration methods. *IEEE transactions on control systems technology*, 18(3):636–653, 2010.
10. Hamed Khorasgani and Gautam Biswas. Structural fault detection and isolation in hybrid systems. *IEEE Transactions on Automation Science and Engineering*, 2017.
11. Abdessamad Kobi, Samuel Nowakowski, and José Ragot. Fault detection-isolation and control reconfiguration. *Mathematics and Computers in Simulation*, 37(2-3):111–117, 1994.
12. Jens Otto, Birgit Vogel-Heuser, and Oliver Niggemann. Automatic parameter estimation for reusable software components of modular and reconfigurable cyber-physical production systems in the domain of discrete manufacturing. *IEEE Transactions on Industrial Informatics*, 14(1):275–282, 2017.

13. Om Prakash and AK Samantaray. Model-based diagnosis and prognosis of hybrid dynamical systems with dynamically updated parameters. In *Bond Graphs for Modelling, Control and Fault Diagnosis of Engineering Systems*, pages 195–232. Springer, 2017.
14. Gregory Provan. Model abstractions for diagnosing hybrid systems. In *Proceedings of the 20th International Workshop on Principles of Diagnosis, DX-09, Stockholm, Sweden*, pages 321–328. Citeseer, 2009.
15. Raymond Reiter. A theory of diagnosis from first principles. *Artificial intelligence*, 32(1):57–95, 1987.
16. Indranil Roychoudhury, Matthew J Daigle, Gautam Biswas, and Xenofon Koutsoukos. Efficient simulation of hybrid systems: A hybrid bond graph approach. *Simulation*, 87(6):467–498, 2011.
17. Daniel Sabin and Eugene C Freuder. Configuration as composite constraint satisfaction. In *Proceedings of the Artificial Intelligence and Manufacturing Research Planning Workshop*, pages 153–161. AAAI Press Palo Alto, CA, 1996.
18. Peter Struss. Fundamentals of model-based diagnosis of dynamic systems. In *IJCAI (1)*, pages 480–485, 1997.
19. Stefan Windmann, Oliver Niggemann, and Heiko Stichweh. Computation of energy efficient driving speeds in conveying systems. *at-Automatisierungstechnik*, 66(4):308–319, 2018.
20. Minjie Zou, Felix Ocker, Edward Huang, Birgit Vogel-Heuser, and Chun-Hung Chen. Design parameter optimization of automated production systems. In *2018 IEEE 14th International Conference on Automation Science and Engineering (CASE)*, pages 359–364. IEEE, 2018.

Machine learning for reconstruction of highly porous structures from FIB-SEM nano-tomographic data*

Chiara Fend[1,2], Ali Moghiseh[2], Claudia Redenbach[1], and Katja Schladitz[2]

[1] Department of Mathematics, University of Kaiserslautern, Germany
`redenbach@mathematik.uni-kl.de`
[2] Fraunhofer-Institut für Techno- und Wirtschaftsmathematik, Kaiserslautern, Germany
{`chiara.fend,ali.moghiseh,katja.schladitz`}`@itwm.fraunhofer.de`

Abstract. Reconstruction of highly porous structures from FIB-SEM image stacks is a difficult segmentation task. Supervised machine learning approaches demand large amounts of labeled data for training, that are hard to get in this case. A way to circumvent this problem is to train on simulated images. Here, we report on segmentation results derived by training a convolutional neural network solely on simulated FIB-SEM image stacks of realizations of a variety of stochastic geometry models.

Keywords: U-net 3D, shine through artifacts, SEM simulation, Boolean model, random packing, Altendorf-Jeulin model, Cox-Boolean model, deep learning

1 Introduction

The micro-structure of materials influences their macroscopic properties decisively. 3D images of the micro-structure yield deeper insight into the micro-structure's geometric features and can be used for numeric simulations of materials properties like mechanical strength, filtration properties or thermal conductivity. Combined with stochastic geometry models [10, 16], they are the basis for optimizing the micro-structure – so-called virtual material design.

Serial slicing by a focused ion beam (FIB) and subsequent imaging by scanning electron microscopy (SEM) is a versatile source for high-quality 3D images of materials structures at the scale of 3-200 nm. For highly porous structures however, reconstruction of the 3D structure from the SEM image stack is hampered by the solid structure from deeper layers being visible through the pores. These so-called shine through artifacts [12] cause the typical tails visible in the planes orthogonal to the SEM imaging plane, see Figure 1(c). These artifacts featuring the same gray values as the true foreground in the current slice, thresholding methods fail to segment the solid structure properly. Several algorithms

---

* Supported by German Federal Ministry of Education and Research, project 03VP00491/5, and Fraunhofer FLAGSHIP PROJECT ML4P.

J. Beyerer et al. (Hrsg.), *Machine Learning for Cyber Physical Systems*, Technologien für die intelligente Automation 13,
https://doi.org/10.1007/978-3-662-62746-4_13

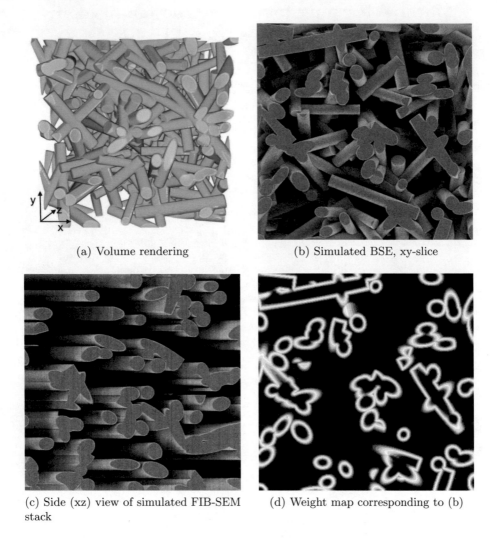

(a) Volume rendering

(b) Simulated BSE, xy-slice

(c) Side (xz) view of simulated FIB-SEM stack

(d) Weight map corresponding to (b)

Fig. 1: Realization of a Boolean model of fibers and synthetic FIB-SEM image stack generated from it. The SEM imaging plane is xy, z is the slicing direction. The xz-view exhibits strong shine through effects.

have been devised to overcome this problem [7, 11, 13, 14, 19, 20]. Nevertheless, being designed for particular structures and SEM modes, they are not generally applicable. Moreover, parameterization requires expert knowledge. Machine learning methods are a popular and already widely used alternative to classical image segmentation. A random forrest is applied to FIB-SEM segmentation in [17]. Convolutional neural networks (CNN) are used with great success also for 3D image segmentation [4, 15]. However, in our particular setting of FIB-SEM data of highly porous structures, the typical need of these methods for large labeled data is nearly prohibitive. FIB-SEM is rather expensive and manual labeling is difficult to impossible as even the human eye is easily mislead by the shine through artifacts.

Here, we therefore explore the option to train a CNN solely based on synthetic images for which the correct segmentation is readily available. We use a variety of stochastic geometry models [1, 8, 10] to create porous structures. Digitizations into 3D images of the respective realizations yield the ground truth for the training phase. The corresponding FIB-SEM stacks are generated based on an analytic representation of the structures – lists of points in space and objects like spheres or cylinders attached to them. These geometries are virtually intersected and SEM images of each planar intersection are simulated as described in [12]. The thus derived FIB-SEM stacks are then used to train the U-net 3D architecture [4, 18].

Specially adapted data augmentation and weights as well as the use of versatile structures result in very good segmentations for the synthetic data. Tests on real data are promising, too, results will be reported in [5].

2   Network architecture and and training the model

We keep the U-net 3D architecture as specified in [4]. We also follow the original U-net setup [18] in using weighted cross-entropy for measuring similarity of image patches. [18] uses weight maps to assign higher weights to pixels in image regions where objects touch, in order to separate them. We adapt this idea by assigning a higher weight to surface pixels and their neighbors in order to force the network to learn the structure's surface particularly well.

The use of 3D patches causes them to be small (64^3 pixels) compared to the total image sizes (about $500^3 - 600^3$ pixels). Simple tiling results in strong boundary artifacts. To avoid these effects, we therefore apply a sliding window approach with up to 20 pixel wide overlapping regions, depending on local structure size.

In [18], data are augmented excessively by deforming the training images elastically, to force the network to learn invariance to such formations. We combine this approach with brightening and rotating the 3D patches using any of the cube's isometries similar to [4]. To ensure that the training data represent various sizes of the local structures, the patches are chosen with a random crop and resize approach, where the scale factor is variable too.

We trained the network on a Boolean model of fibers (cylinders) [21, 22] as, compared to models consisting solely of spheres, the cylinders yield a wider variety of local structures, e. g. both circular and very elongated elliptical cross-sections. More precisely, we used a Boolean model of straight cylinders with circular cross-section of diameters uniformly distributed in $60 - 90$ nm and with lengths uniformly distributed in $300 - 660$ nm. The orientations are uniformly distributed on the upper half-sphere. That is, the model is isotropic. The solid volume fraction is 35%. See Figure 1 for the model realization used.

From the model realization, the FIB-SEM stack is simulated based on [9] as described in [12]. Throughout, the back scattered electron (BSE) signal is used. The solid component is assumed to be carbon, the primary electrons have an energy of 5 keV as in [12], and the dwelling time is 1μs (see e. g. [6] for details on SEM parameters). Both SEM image pixel size and slice distance are 3 nm. The training lasted for 100 epochs, where one epoch equals 50 steps with a batch size of 4. The initial learning rate of 0.0001 is halved after every 10 epochs.

3 Results

The network trained as described in the previous section is now used to segment synthetic FIB-SEM image stacks of a variety of other structures. More precisely, we segment images of Boolean models of spheres and a Cox Boolean model [8] of small spheres nested in large ones (see Figure 2(a)). These are complemented by packings of spheres by the force biased algorithm [2, 3], of straight circular cylinders by random sequential adsorption (RSA), and of curved fibers by the Altendorf-Jeulin method (AJ) [1].

We measure the quality of our results by the false negative rate (FNR, the proportion of missed foreground pixels) and the Sørensen-Dice coefficient [23]. The latter is defined as

$$DICE(\hat{y}, y) = \frac{2 * \sum_{i=1}^{n} \hat{y}^i * y^i + \varepsilon}{\sum_{i=1}^{n} \hat{y}^i + \sum_{i=1}^{n} y^i + \varepsilon} = \frac{2 * TP + \varepsilon}{FN + 2 * TP + FP + \varepsilon},$$

where n is the total number of pixels in the volume, y represents the pixel-wise ground truth, $\hat{y}$ the predicted (segmented) image, and TP, TN, FP, FN are counts of true positive, true negative, false positive and false negative pixels in the prediction, respectively.

Due to the shine through effects described above and the typical coarser sampling in slicing direction, structures reconstructed from FIB-SEM image data tend to be anisotropic to an extent not explainable by the sample production or preparation [14]. All structures considered here are isotropic by design. That is, the distributions of the respective stochastic geometry models are invariant under rotations. Isotropy of the reconstructed structures is therefore a measure for their quality, too. It is by far not an easy task to test the realization of a random closed set for isotropy. Here, we just check the proportion of the mean chord lengths in x- and z-directions $\bar{\ell}_x$ and $\bar{\ell}_z$ as a rough indicator of artificial anisotropy. The only suspicious case is the RSA cylinder packing. However, here already the 3D ground truth image has a mean chord length ratio of $\bar{\ell}_x/\bar{\ell}_z = 0.88$.

All results are listed in Table 1 and visualized in Figure 2.

4 Conclusion

In this contribution, we show that a deep neural network – namely an adapted U-Net 3D – trained solely on synthetic FIB-SEM image stacks, is capable to reconstruct other highly porous structures from FIB-SEM images.

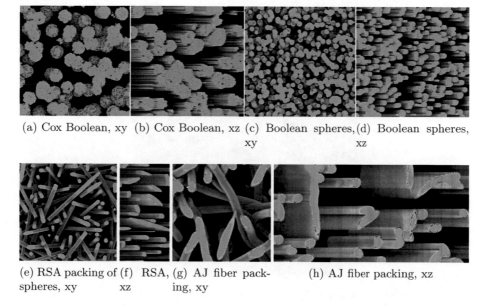

(a) Cox Boolean, xy (b) Cox Boolean, xz (c) Boolean spheres,(d) Boolean spheres,
xy   xz

(e) RSA packing of (f) RSA, (g) AJ fiber pack-   (h) AJ fiber packing, xz
spheres, xy   xz   ing, xy

Fig. 2: Segmentation results visualized as overlays on the original FIB-SEM images. Green pixels are correctly segmented foreground pixels (true positive), pixels misclassified as being foreground (false positive) are represented in yellow,
and the ones that are misclassified as being background are colored red (false
negative).

Table 1: Results of the network trained using solely the Boolean cylinder model
as shown in Figure 1.

	Data					
	Boolean model of fibers	Boolean model of spheres	RSA packing of cylinders	Cox Boolean model of spheres	Altendorf-Jeulin fiber packing	Forced biased packing of spheres
DICE	0.95	0.95	0.96	0.96	0.95	0.99
FNR	0.01	0.01	0.05	0.07	0.05	0.01
$\bar{\ell}_x/\bar{\ell}_z$	0.98	1.02	0.80	0.94	0.99	0.91

More details on the stochastic geometry models, wider variation of the FIB-
SEM imaging parameters, and results on real data will be presented in [5].

The experiments leading to the results presented here highlighted the need for
sufficient diversity of local structures in the data used for training. As a caution-
ary example we show in Figure 3 a result obtained for a cylinder packing using
exactly the same network as described above, but trained solely on a Boolean

model of spheres. Clearly, the thus "mis-trained" net tries to approximate the foreground by spheres.

Fig. 3: Overlaid result for RSA fiber packing obtained with net trained on Boolean model of spheres. Meaning of colors as in Figure 2 above.

References

1. Altendorf, H., Jeulin, D.: Random-walk-based stochastic modeling of three-dimensional fiber systems. Phys. Rev. E **83**, 041804 (Apr 2011)
2. Bezrukov, A., Bargieł, M., Stoyan, D.: Statistical analysis of simulated random packings of spheres. Part. Part. Systems Char. **19**, 111–118 (2002)
3. Bezrukov, A., Stoyan, D., Bargieł, M.: Spatial statistics for simulated packings of spheres. Image Anal. Stereol. **20**, 203–206 (2001)
4. Çiçek, Ö., Abdulkadir, A., Lienkamp, S.S., Brox, T., Ronneberger, O.: 3D U-Net: learning dense volumetric segmentation from sparse annotation. In: International Conference on Medical image computing and computer-assisted intervention. pp. 424–432. Springer (2016)
5. Fend, C., Moghiseh, A., Redenbach, C., Schladitz, K.: Reconstruction of highly porous structures from FIB-SEM using a deep neural network trained on synthetic images (2019), in preparation.
6. Goldstein, J., Newbury, D., Joy, D., Lyman, C., Echlin, P., Lifshin, E., Sawyer, L., Michael, J.: Scanning electron microscopy and X-ray microanalysis, 3rd edition. Springer science, New York (2003)
7. Holzer, L., Stenzel, O., Pecho, O., Ott, T., Boiger, G., Gorbar, M., de Hazan, Y., Penner, D., Schneider, I., Cervera, R., Gasser, P.: Fundamental relationships between 3d pore topology, electrolyte conduction and flow properties: Towards knowledge-based design of ceramic diaphragms for sensor applications. Materials and Design **99**, 314 – 327 (2016)

8. Jeulin, D.: Morphology and effective properties of multi-scale random sets: A review. Comptes Rendus Mécanique **340**(4), 219 – 229 (2012), Recent Advances in Micromechanics of Materials

9. Lowney, J.R.: Monte Carlo simulation of scanning electron microscope signals for lithographic metrology. Scanning **18**(4), 301–306 (1996)

10. Ohser, J., Schladitz, K.: 3d Images of Materials Structures – Processing and Analysis. Wiley VCH, Weinheim (2009)

11. Penner, D., Holzer, L.: Characterization and modelling of structure and transport properties of porous ceramics. Publikationen School of Engineering: ZHAW Digital Collection **95**(3), E27–E32 (2018)

12. Prill, T., Schladitz, K.: Simulation of FIB-SEM images for analysis of porous microstructures. Scanning **35**, 189–195 (2013)

13. Prill, T., Schladitz, K., Jeulin, D., Faessel, M., Wieser, C.: Morphological segmentation of FIB-SEM data of highly porous media. Journal of Microscopy **250**(2), 77–87 (2013)

14. Prill, T., Redenbach, C., Roldan, D., Godehardt, M., Schladitz, K., Höhn, S., Sempf, K.: Simulating permeabilities based on 3d image data of a layered nanoporous membrane. International Journal of Solids and Structures (2019)

15. Qi, C.R., Su, H., Mo, K., Guibas, L.J.: Pointnet: Deep learning on point sets for 3d classification and segmentation. In: Proceedings of the IEEE Conference on Computer Vision and Pattern Recognition. pp. 652–660 (2017)

16. Redenbach, C., Schladitz, K., Vecchio, I., Wirjadi, O.: Image analysis for microstructures based on stochastic models. GAMM-Mitteilungen **37**(2), 281–305 (2014)

17. Röding, M., Fager, C., Olsson, A., von Corswant, C., Olsson, E., Lorén, N.: Three-dimensional reconstruction of microporous polymer films from FIB-SEM nanotomography data using random forests. Microscopy & Micoranalysis (2019), submitted.

18. Ronneberger, O., Fischer, P., Brox, T.: U-Net: Convolutional networks for biomedical image segmentation. In: International Conference on Medical image computing and computer-assisted intervention. pp. 234–241. Springer (2015)

19. Salzer, M., Prill, T., Spettl, A., Jeulin, D., Schladitz, K., Schmidt, V.: Quantitative comparison of segmentation algorithms for FIB-SEM images of porous media. Journal of Microscopy **257**(1), 23–30 (2015)

20. Salzer, M., Thiele, S., Zengerle, R., Schmidt, V.: On the importance of FIB-SEM specific segmentation algorithms for porous media. Materials Characterization **95**, 36 – 43 (2014)

21. Schneider, R., Weil, W.: Stochastic and Integral Geometry. Probability and Its Applications, Springer, Heidelberg (2008)

22. Stoyan, D., Kendall, W.S., Mecke, J.: Stochastic Geometry and Its Applications. Wiley, Chichester, 2nd edn. (1995)

23. Taha, A.A., Hanbury, A.: Metrics for evaluating 3D medical image segmentation: analysis, selection, and tool. BMC medical imaging **15**, 29 (Aug 2015)

Printed in the United States
By Bookmasters